Advances in Microfluidic Technologies for Energy and Environmental Applications

Edited by Yong Ren

Published in London, United Kingdom

IntechOpen

Supporting open minds since 2005

Advances in Microfluidic Technologies for Energy and Environmental Applications
http://dx.doi.org/10.5772/intechopen.81935
Edited by Yong Ren

Contributors
Jing Wang, Bei Zhang, Yong Ren, Voon Loong Wong, Chin-Ang Isaac Ng, Lui-Ruen Irene Teo, Ci-Wei Lee, Tran Long, Yehezkiel Steven Kurniawan, Keisuke Ohto, Ramachandra Rao Sathuluri, Rui Lima, Inês Maia, Cesar Rocha, Pedro Pontes, Vanessa Cardoso, J. M. Miranda1, Ana S. Moita, G. Minas, António L. N. Moreira, Maxine Yew, Kai Seng Koh, Faik Hamad, Min Yang, Changhe Li

First published in London, United Kingdom, 2020 by IntechOpen
IntechOpen is the global imprint of INTECHOPEN LIMITED, registered in England and Wales, registration number: 11086078, 7th floor, 10 Lower Thames Street, London, EC3R 6AF, United Kingdom
Printed in Croatia

British Library Cataloguing-in-Publication Data
A catalogue record for this book is available from the British Library

Additional hard and PDF copies can be obtained from orders@intechopen.com

Advances in Microfluidic Technologies for Energy and Environmental Applications
Edited by Yong Ren
p. cm.
Print ISBN 978-1-78984-418-4
Online ISBN 978-1-78984-419-1
eBook (PDF) ISBN 978-1-83880-545-6

An electronic version of this book is freely available, thanks to the support of libraries working with Knowledge Unlatched. KU is a collaborative initiative designed to make high quality books Open Access for the public good. More information about the initiative and links to the Open Access version can be found at www.knowledgeunlatched.org

For EU product safety concerns:
IN TECH d.o.o., Prolaz Marije Krucifikse Kozulić 3, 51000 Rijeka, Croatia, info@intechopen.com or visit our website at intechopen.com.

Meet the editor

Dr. Yong Ren received his PhD degree in mechanical engineering from Hong Kong Polytechnic University. He is an Associate Professor in the Department of Mechanical, Materials, and Manufacturing Engineering at the University of Nottingham Ningbo China, and Director of the Ningbo Key Laboratory on Energy Material and Technology. He has a broad range of research interests including micro-/nano-scale fluid dynamics, heat and mass transfer, multiphase flow, multi-field coupling problems, and bio-MEMS applications. His present research focuses on development of functional micro/nano materials using microfluidics for environmental applications.

Contents

Preface

This book aims to provide a comprehensive review of the latest advances in applying the microfluidic platform and technologies for addressing the problems and challenges arising from environmental and energy areas. Nine chapters are included in the book, covering a wide range of topics and establishing a basic framework that will be helpful for readers to appreciate the important role that has been played by microfluidics and envision the new areas and problems that deserve more efforts to explore in future endeavors.

It is the outstanding contribution from the academic researchers working in the fields that make this book possible. The authors of the book chapters are greatly acknowledged, and the sincere gratitude also goes to the publisher.

Yong Ren
University of Nottingham Ningbo China,
Ningbo, China

Section 1

Environmental and Biomedical Applications

Chapter 1

Droplet Microfluidic Device for Rapid and Efficient Metals Separation Using Host-Guest Chemistry

Yehezkiel Steven Kurniawan, Ramachandra Rao Sathuluri and Keisuke Ohto

Abstract

Metals are pivotal elements in our daily life and industrial processes, to produce electronic devices, catalysts, smart materials and so on. However, they are mostly present as a mixture in the environment that makes their separation challenging over the past decade. Host-guest chemistry principle thoroughly has been used to design and synthesize thousands of organic receptors with high complexation ability and selectivity to certain metal ions. On the other hand, the droplet microfluidic device is well-known for its unique characteristics of fluid dynamics, such as large specific surface area and short diffusion distance making the process robust and efficient. Therefore, many reports of research employ host-guest chemistry of the droplet microfluidic system for the effective metal separation process. This chapter deals with up-to-date examples of the droplet microfluidic system application for separation of base and alkali metals, recovery of rare-earth and precious metals and removal of heavy metals either from the competitive metal system or from the real waste solution sample through solvent extraction techniques utilizing host-guest chemistry principle.

Keywords: droplet microfluidic device, host-guest chemistry, solvent extraction, recovery, metal separation

1. Introduction

Nowadays, smart and advanced technology researches have infiltrated on all aspects of human daily activities [1]. For example, smart devices containing metal elements have massively been applied in pharmaceutical and electronic fields [2]. In general, based on their properties and usage, metal elements can be divided into four groups, *i.e.* alkali and base metals (lithium, iron, cobalt, nickel, zinc, etc.), precious metals (silver, gold, palladium, platinum, rhodium, iridium, osmium and ruthenium), rare-earth metals (lanthanum, neodymium, europium, holmium, etc.) and heavy metals (cadmium, lead, mercury, chromium, etc.). Silver materials have been applied as an antimicrobial agent while lithium ones as well-known energy storage materials [3]. The demands for these smart materials have kept increasing, therefore, their supply has been depleting day-by-day. Global maldistribution of metals, furthermore, is worsened by the metal supplies in several countries. Over the past several years, researchers are putting their continuous efforts on metal recycling and recovery process from wastewater and other secondary resources [4].

Metal recycling is a process to recover metals either from generated solid waste/liquid metal waste or spent home appliances using several techniques for a constant supply to meet the demand. Many kinds of research have been conducted to evaluate the metal recycling process; however, the metal recycling is quite complicated due to the presence of other metals with similar physicochemical properties in varying concentrations [1]. Additionally, heavy metals removal from the environment is difficult because they usually exist in trace amounts compared with other elements [5]. Therefore, versatile techniques which exhibit high metal selectivity and high recovery percentage are highly required [6].

Solvent extraction or liquid-liquid extraction is one of the metal separation techniques where the metal extraction efficiency mainly depends on the nature of the extraction reagents used. For example, di(2-ethylhexyl)phosphoric acid (D2EHPA) extracted lead(II) from chloride aqueous solution, however, the selectivity against iron(III), zinc(II) and copper(II) ions is unsatisfied [7]. Meanwhile, a complete separation of lead(II) ions from iron(III), zinc(II) and copper(II) successfully achieved using an amide derivative of calix[4]arene due to its high and suitable host-guest interaction [8].

On the other hand, droplet microfluidic devices are popular in biology and analytical chemistry areas because of their unique characteristics and advantages [9–15]. Numerous research papers related to rapid and efficient metal separations reported recently, and their numbers keep increasing with time [16–20]. This chapter serves an up-to-date review on the application of the droplet microfluidic system for separation and recovery of metal elements using a combination of host-guest chemistry principle and solvent extraction technique.

2. Host-guest chemistry

Host-guest chemistry is one of the chemistry fields studies various non-covalent interactions between host and guest molecules [21]. Host molecule defined as a molecular entity that forms complexes with organic or inorganic guests, or a chemical species that can accommodate guests within cavities of its crystal structure. The guest molecule is a molecule having a complementary structure to the binding site of the host compound, thus generating a selective and strong interaction between them [22]. The non-covalent interactions between host and guest molecules are including size effect, hydrophobic, cation-π, π-π, electrostatic, and hard-soft acid-base interactions as well as hydrogen bonding. Because of these specific interactions, a selective complexation between host and guest molecules could be achieved [23].

Thousands of host molecules are being prepared and evaluated to find out the best suitable host candidate for metal separations. Among them, calixarenes are found to be interesting host molecules and they attracted a lot of attention for research works due to their special inclusion phenomena [24]. Calixarenes are macrocyclic compounds that are prepared from formaldehyde and *p*-alkylphenol as the repeating units. A single step preparation of calixarenes was firstly introduced by Prof. C.D. Gutsche in 1978, subsequently, it has grown rapidly due to some merits, such as large-scale synthesis, rigid structure, easy modification and stable host compound [25]. It is also interesting to note that, the ring size and conformation of calixarenes are controlled by using template and reaction temperature.

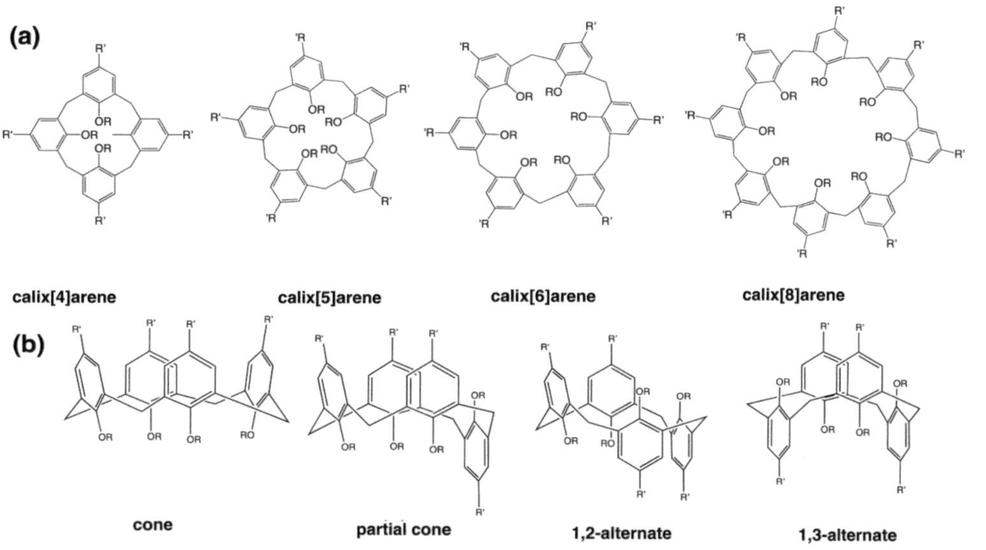

Figure 1.
(a) The structures of calix[n]arenes and (b) the conformations of calix[4]arene.

So far, there are four main subfamilies of calixarenes based on their ring size, *i.e.* calix[4]arene, calix[5]arene, calix[6]arene and calix[8]arene as displayed in **Figure 1(a)**. By increasing the ring size of calixarenes, their conformation could not easily be adjusted because of weaker hydrogen bonding on the calixarene structure [26].

Calix[4]arene is the most simple and special subfamily of calixarenes because it serves the most stable conformations. In general, there are four available conformations of calix[4]arene, *i.e.* the cone, partial cone, 1,2-alternate and 1,3-alternate conformations as shown in **Figure 1(b)**. Calix[4]arenes offer ideal frameworks for metal separation because of their size discrimination and chelating effects that contributed to high selectivity and high complexation ability with certain metal ions [27]. Furthermore, it is possible to change its sequence of metal selectivity by modifying their upper and lower rims [28]. Therefore, it is reasonable to understand that a lot of research works utilized calix[4]arenes for metal extraction, metal adsorption, and as chemosensor of metal ions, etc. [29–31].

3. Droplet microfluidic device

Microfluidic system is a miniaturized device that manipulates and controls fluids, usually in the range of microliters to femtoliters, in the channels with dimensions from tens to hundreds of micrometers [14]. The earliest microfluidic system was reported by Hodgson and Charles in 1963 while they investigated droplet generation when oil and water phases were introduced into the microchannel [32]. In 2001, Burns and Ramshaw firstly reported that the mass transfer of acetic acid from kerosene to aqueous phase was significantly enhanced by using droplet microfluidic device [33]. As of today, there are many types of microfluidic devices, such as continuous microfluidic devices, droplet microfluidic devices, paper microfluidic devices, and digital microfluidic devices have been fabricated and applied for many applications [34, 35]. Each of them serves as a unique platform for certain applications. Continuous microfluidic devices have been widely applied to particles and cell separation due to density differences. Droplet microfluidic devices serve as an ideal platform for metal separation, nanoparticle fabrication, drug delivery, and cell assay. Paper microfluidic devices have also been reported for routine chemical analysis and biochemical assays, while digital microfluidic devices are employed for immunoassays and chemical synthesis, etc. [36].

As the name indicates, the droplet microfluidic device generates droplets of two immiscible liquids such as organic and aqueous phases in the microchannel. Based on the liquids' intersection, there are 2 types of droplet microfluidic devices, *i.e.* T-type and Y-type of droplet microfluidic devices [9, 12]. A schematic representation of a droplet microfluidic device is shown in **Figure 2**. When the liquid is introduced into the microchannel through an inlet which is perpendicular to the main microchannel of a T-type droplet microfluidic device, it was defined as the dispersed phase, while, the continuous phase liquid that has introduced through an inlet which is in line with the main microchannel of a T-type droplet microfluidic device. However, both phases are equal to the Y-type of droplet microfluidic devices [13].

Kashid *et al.* have observed the droplets generation on both T-type and Y-type droplet microfluidic devices in 2010 [37]. For example, on the T-type droplet microfluidic device, first, the dispersed phase is entered the main microchannel and starting to block the flow of the continuous phase by exerting the pressure. The generated pressure squeezes the neck of the dispersed phase and the droplet of the dispersed phase is formed. Since both phases are immiscible, each droplet has a role as an individual reactor with a short diffusion distance. Furthermore, each droplet serves a large specific surface area. The combination of a short diffusion distance and a large specific surface area contribute to boosting the mass transfer process of chemicals from one phase to another phase [38].

Even though solvent extraction offers a selective and efficient technique for metal separation, the extraction kinetics rate in some processes is slow and time-consuming [38]. For comparison, **Figure 3** shows the obtained parameters and experimental data for silver(I) extraction using a tetramethylketonic derivative of calix[4]arene in batch-wise and microfluidic systems. In the batch-wise system, the extraction was carried out in a small glass vial with a diameter of 4 cm and height of each phase is about 1 cm. The extraction using a continuous microfluidic device was performed in a microchannel with 20 mm of length and 200 μm in depth and width while the extraction in a droplet microfluidic device was carried out in a microchannel with 73 mm of length and 200 μm of depth and widths. As shown in **Figure 3**, the diffusion distance of chemicals using batch-wise (20,000 μm) was a hundred times farther than both the microfluidic devices (200 μm). Furthermore, the total specific surface area of the droplet microfluidic device employed was ~550 times larger than that of the continuous microfluidic system while ~55,000 times larger than that of the batch-wise system. Because of that, it is reasonable to know that the required extraction time to reach an equilibrium state is 259,200 s (72 h), 15.0 s and 1.00 s for the batch-wise system, continuous microfluidic system, and droplet microfluidic system, respectively. This result demonstrates the remarkable advantage of using the droplet microfluidic device for metal extraction and separation process [39].

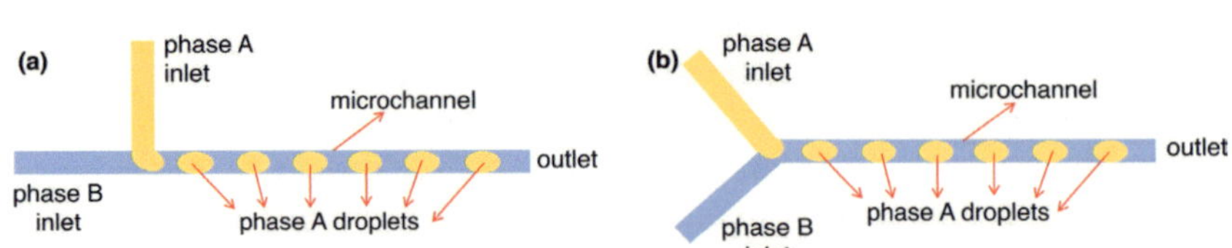

Figure 2.
(a) T-type and (b) Y-type of droplet microfluidic devices.

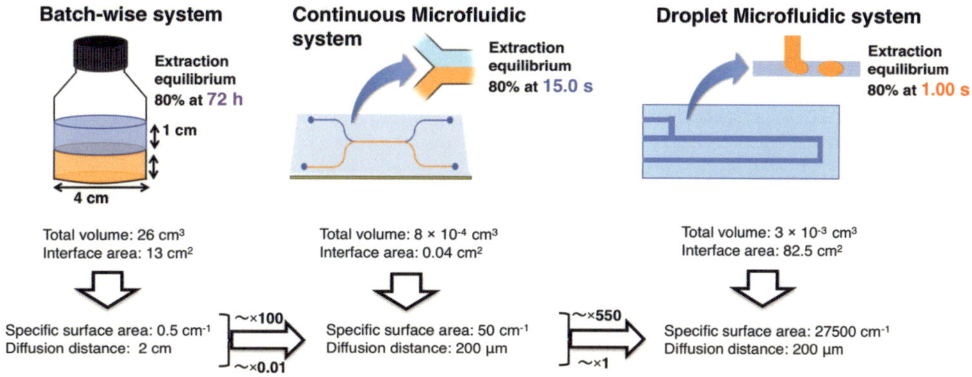

Figure 3.
Comparison of batch-wise, continuous and droplet microfluidic systems for silver(I) extraction using a tetramethylketonic derivative of calix[4]arene [38].

4. Metal separations using the droplet microfluidic device

Due to the significant enhancement of the extraction process, droplet microfluidic devices have been employed for metal ion separations through a solvent extraction technique [18–20]. The chemical structures of several extraction reagents employed in metal separations are shown in **Figure 4**. In general, metal ion concentrations were measured using either atomic absorption spectroscopy (AAS) or induced coupled plasma-atomic emission spectroscopy (ICP-AES) or induced coupled plasma-mass spectrometry (ICP-MS). The measured metal ion concentrations were further used to calculate extraction percentages (%E), distribution ratio (D) and separation factor (β) as mentioned in Eqs. (1)–(3) whereas $[metal\ ion]_{aqueous,0}$ is initial metal ion concentration in the aqueous phase before solvent extraction process, $[metal\ ion]_{aqueous,t}$ is the metal ion concentration in the aqueous phase after t extraction time, and $[metal\ ion]_{organic,t}$ is the metal ion concentration in the organic phase after t extraction time [26]. The metal extraction and separation processes are summarized and listed in **Table 1**.

$$\%Extraction = \frac{[metal\ ion]_{aqueous,0} - [metal\ ion]_{aqueous,t}}{[metal\ ion]_{aquoeus,0}} \times 100 \quad (1)$$

$$D = \frac{[metal\ ion]_{organic,t}}{[metal\ ion]_{aqueous,t}} \quad (2)$$

$$\beta = \frac{D_{metal\ ion\ A}}{D_{metal\ ion\ B}} \quad (3)$$

4.1 Alkali and base metals

Alkali and base metals are non-precious metals which are abundantly found in nature [40]. In this part, the example of utilization of droplet microfluidic system for extraction and separation of copper(II), iron(III), indium(III), calcium(II), cesium and lithium ions were discussed.

Evaluation of copper(II) extraction using Y-type microreactor was carried out by Yang *et al*. The commercially available extractant, AD-100 which consists of 70–75% (w/w) of 2-hydroxy-5-nonylbenzaldehyde oxime as the active reagent in 260# solvent oil as the organic diluent was used. The concentrations of AD-100 used influenced the copper(II) extraction percentage. Higher AD-100

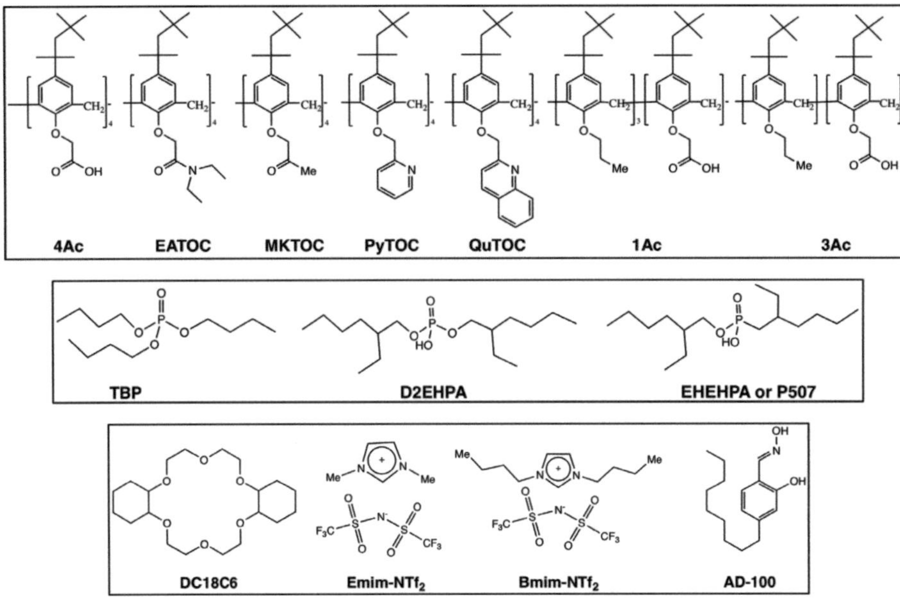

Figure 4.
Host compounds and extraction reagents used for metal separations in this chapter.

concentration gave higher extraction percentage while a higher concentration of copper(II) gave lower extraction percentage. The extraction percentages were also increased at longer extraction time, lower Reynolds number, higher pH of the aqueous phase, and higher temperature. Using the T-type of droplet microfluidic device, 96.0% copper(II) ions were extracted in 32.0 s extraction time [41]. On the other hand, Jiang *et al*. employed a Y-type microfluidic device to achieve an effective separation of copper(II) over iron(III) and zinc(II) with a mixture of 2-hydroxyl-5-nonyl acetophenoneoxime and 5-nonylsalicyladoxime in 1:1 volume ratio, called as DZ988N, as the extraction reagent. Using the microfluidic device, as much as 80.6% of copper(II) extracted at 0.78 s extraction time. While as much as 98.9% of copper(II) was stripped at 3.92 s by using 1.84 M of sulfuric acid solution in 2:1 organic to aqueous phase ratio. In a comparison of the microfluidic device to the batch-wise system, the separation factors of copper(II) over iron(III) and zinc(II) were significantly enhanced from 86.2 to 466 and 123 to 1089, respectively [42].

Iron(III) extraction kinetics using D2EHPA was investigated in a Y-type microfluidic device. The experimental results showed that the extraction of iron(III) was controlled by mixed reaction-diffusion resistance regime on a second-order extraction rate equation. Moreover, since the required extraction times for iron(III) and cobalt(II) are different, iron(III) could be separated from the cobalt(II) in the metal mixture solution, which may contribute to the real hydrometallurgical process [43].

Indium(III) separation from a metal mixture solution has been successfully achieved using a T-type of droplet microfluidic device. The metal mixture contains 23.0 mM of indium(III), 35.8 mM of iron(III), 44.2 mM of zinc(II), 153.8 mM of aluminum(III) and 141.5 mM of magnesium(II) ions [44]. It was found that indium(III) was extracted in 98.8%, while the other metal ions were hardly extracted (<5.00%). The separation process yielding that the separation factors of indium(III) over iron(III), zinc(II), aluminum(III) and magnesium(II) ions were 18,720, 266,373, 41,955 and 11,809, respectively, demonstrating that a high selective indium(III) separation was achieved within a minute.

Sl. No	Metal ions	Extraction reagents	Microchannel			%*E*	β	*t* (s)	Ref
			Length (mm)	Width (mm)	Type				
Base metals									
1	Cu(II)	AD-100	900	0.6	T	96.0	—	32.0	[41]
2	Cu(II)	DZ988N	131	0.2	Y	80.6	$466^{Fe(III)}$ $1250^{Zn(II)}$	0.78	[42]
3	Fe(III)	D2EHPA	3	0.1	Y	32.0	—	—	[43]
4	Co(II)	D2EHPA-Na	500	1.0	Y	92.0	$3.54^{Li(I)}$	15.0	[50]
5	In(III)	D2EHPA	2000	0.8	T	98.8	$18720^{Fe(III)}$ $266373^{Zn(II)}$ $41955^{Al(III)}$ $11809^{Mg(II)}$	60.0	[44]
Alkali metals									
6	Ca(II)	DC18C6	20	0.5	Y	98.2	—	5.70	[45]
7	Ca(II)	$EMIMNTf_2$ + $BMIMNTf_2$	130	0.4	Y	52.0	—	6.00	[46]
8	Cs(I)	D2EHPA	22	0.7	Y	100	—	40.0	[47]
9	Li(I)	D2EHPA + TBP	500	1.0	Y	50.0	—	5.00	[49]
10	Li(I)	1Ac	73	0.2	T	100	> 1000	4.00	[51]
Precious metals									
11	Ag(I)	MKTOC	73	0.2	T	100	—	4.00	[52]
12	Pd(II)	QuTOC	73	0.2	T	100	—	4.00	[52]
13	Pt(IV)	PyTOC	73	0.2	T	100	—	4.00	[52]
Rare-earth metals									
14	La(III)	P507	120	0.1	Y	99.5	—	0.37	[55]
15	La(III)	Phosphorus reagent	100	1.0	T	90.0	$3.50^{Eu(III)}$	10.0	[56]

Sl. No	Metal ions	Extraction reagents	Microchannel			%E	β	t (s)	Ref
			Length (mm)	Width (mm)	Type				
16	Nd(III)	P507	125	0.3–1.0	Y	88.0	$2.23^{Pr(III)}$	12.0	[58]
17	Ce(III)	P507	125	0.3	Y	76.5	$3.25^{Pr(III)}$	12.0	[57]
18	Sm(III)	P507	125	0.3	T	75.1	—	45.0	[59]
19	U(IV)	TBP	13	0.5	T	100	—	1.50	[65]
Heavy metals									
20	Pb(II)	4Ac	73	0.2	T	96.0	$14.2^{Fe(III)}$ $324^{Cu(II)}$ $296^{Zn(II)}$	4.00	[66]
21	Pb(II)	EATOC	73	0.2	T	100	> 1000	2.00	[8]

Table 1.
Summary of reports on metal ion extractions and separations carried out in the microfluidic devices at t *extraction time.*

Jahromi *et al.* separated calcium(II) isotopes, ^{48}calcium(II) and ^{40}calcium(II) ions using dicyclohexano-18-crown-6 in a Y-type microfluidic device. Using a combination of the dicyclohexano-18-crown-6 and microfluidic device, ^{40}calcium(II) was selectively extracted in 98.2% within 5.70 s. By using a continuous cycle extraction process, the amount of ^{40}calcium(II) becomes lower in the aqueous phase, therefore, the concentration of ^{48}calcium(II) ion was enriched [45]. Another experiment on calcium(II) extraction was carried out by using ionic liquids as the extraction reagent, named as 1-ethyl-3-methylimidazolium bis(trifluoromethylsulfonyl)imide (EMIM NTf2) and 1-butyl-3-methylimidazolium bis(trifluoromethylsulfonyl)imide (BMIM NTf2) in a Y-type of droplet microfluidic device. The pH value of the aqueous phase did not affect the extraction percentage because the main extraction mechanism is not an ion-exchange mechanism. Even though some parameters were optimized, the highest extraction efficiency was only 52.0%, which was still unsatisfied [46].

On the other hand, cesium(I) extraction in a Y-type of droplet microfluidic system was studied by Tamagawa *et al*. Stable droplet generation was observed when the total flow rate of both phases is less than 100 mL h^{-1}. By using D2EHPA as the extraction reagent, the required times to reach the quantitative amount of extraction percentage (100%) were shortened 25 times from 1000 to 40.0 s using batch-wise and droplet microfluidic system, respectively [47].

Lithium metals are widely used as the main part of lithium-ion batteries and aircraft alloys, the continuous supply of lithium may become critical over time. Urban waste, especially from lithium-ion batteries and seawater, may serve as the potential secondary resource in the future. However, lithium-ion batteries contain lithium, cobalt(II) and manganese(II) ions, while seawater contains lithium and other metal ions, such as sodium, potassium, magnesium, and calcium ions in different concentrations, the lithium recovery from these resources poses to be a tough work [48].

Muto and co-workers investigated the extraction process of lithium using D2EHPA in a droplet microfluidic device. They found that a glass substrate gave higher extraction efficiency as well as the mass transfer coefficient due to stronger convection rate and larger specific surface area of the organic droplets. When 2.0 M D2EHPA was used as the extraction reagent for 1.0 mM of lithium ions in the aqueous phase, around 0.3 mM of lithium ions were extracted in 5.00 s as the extraction time. Furthermore, by the addition of 0.18 M tributyl phosphate (TBP), the amount of the extracted lithium can be improved to 0.5 mM at 5.00 s extraction time. Even though the extraction efficiency was only around 50%, the droplet microfluidic device serves as a potential platform for quick metal extraction compared with the conventional extraction process [49].

Hirayama *et al.* evaluated the separation of cobalt(II) over lithium from a model solution of lithium-ion battery waste. The mixture of 0.20 M of D2EHPA and 0.01 M of NaOH, called as D2EHPA-Na in cyclohexane was used as the organic phase while the model solution containing 1.0 mM of cobalt(II) and 1.0 mM of lithium ions was used as the aqueous phase. From the model solution, the extraction reagent extracted around 50 ppm of cobalt(II) and 1.5 ppm of lithium ions giving 3.54 as the separation factor for cobalt(II) over lithium ions. Furthermore, using the droplet microfluidic device, 98% purity of cobalt(II) solution was obtained within 4.50 s extraction time [50].

In our previous study, we investigated the possibility to extract lithium ions with monoacetic acid derivative of calix[4]arene using T-type of droplet microfluidic system [51]. At first, the droplet microfluidic device was fabricated from Tempax glass and the microchannel was produced utilizing a diamond-coated needle through a micromilling process. The microchannel dimensions were 73.0 mm in length and 200 μm in width and depth as confirmed by laser reflection microscope. The top and bottom parts of the T-type droplet microfluidic device were permanently bonds through a thermal fusion technique [52]. To prevent any disturbance of leaked sodium ions from the glass substrate, the microchannel was coated with SFE-X008 coating agent (1% of *m*-bis(trifluoromethyl)benzene in 94% of the ethyl perfluorobutyl ether). After coating, the wettability of the glass substrate was significantly changed from 20 to 120°, demonstrating a successful coating of the microchannel. From the preliminary investigation on lithium extraction using the monoacetic acid derivative of calix[4]arene (shown in **Figure 4** as 1 Ac) on the individual metal extraction, it was found that 100% of lithium ions were quantitatively extracted just within 2.00 s as the extraction time. Moreover, the complete stripping of lithium ions from the organic phase was easily achieved using 0.10 M HCl as the stripping agent.

Since the monoacetic acid derivative of calix[4]arene completely recovered (extraction and stripping efficiencies were 100%) the lithium ions, the competitive metal extraction was carried out using a mixture of metal ions as the aqueous phase. In a mixture of all alkali and alkaline earth metals (lithium, sodium, potassium, rubidium, cesium, magnesium, calcium, strontium, and barium ions) in 10.0 mM concentration each, the 5.00 mM of the monoacetic acid derivative of calix[4]arene extracted 4.99 mM of lithium ions showing a quantitative (99.8%) extraction percentage of lithium ions. By using 1.00 M of HCl, as much as 99.5% of lithium ions were successfully stripped from the ligand, which is remarkable. The results demonstrated that the lithium separation from seawater which mainly contains sodium, potassium, magnesium and calcium ions was successfully achieved using the monoacetic acid derivative of calix[4]arene in the droplet microfluidic system.

Seawater was sampled from Minatohama, Karatsu, Japan and filtered using Millipore membrane to prevent any blockage on the microchannel during the extraction process. The lithium concentration in the seawater was 0.03 mM while the sodium, potassium, magnesium and calcium ions concentration was 495, 10.1, 47.4 and 12.8 mM, respectively. Using a droplet microfluidic system, it was found that 31.6% of lithium (0.01 mM) and 1.00% of sodium ions (4.95 mM) were extracted from seawater, however, other metal ions were not extracted at all and remained in the aqueous phase. In this case, sodium ions were extracted due to the huge amount of existing sodium (~19,000 times higher) compared to lithium ions.

To achieve complete extraction of lithium ions from seawater, 20.0 mM of the monoacetic acid derivative of calix[4]arene in chloroform was used as the extraction reagent. As expected, 100% of lithium ions were successfully extracted. However, since the higher concentration of the extraction reagent was used, as much as 2.56% of sodium ions were also extracted. Both metal ions were completely stripped using HCl to obtain an aqueous phase containing 0.03 mM of lithium and 12.2 mM of sodium ions. To obtain a pure lithium solution, another type of calix[4]arene, named as triacetic acid derivative of calix[4]arene was used because it exhibits high sodium selectivity over lithium ions at the acidic condition. When the 10.0 mM of triacetic acid derivative of calix[4]arene (shown in **Figure 4** as 3Ac) was employed, 100% of the remained sodium ions were completely extracted leaving pure lithium ions in the aqueous phase. Similar to monoacetic acid derivative, triacetic acid derivative of calix[4]arene was also recycled. The scheme of recovery of lithium ions from seawater is shown in **Figure 5(a)**. From this experiment, a rapid and efficient lithium-ion recovery from the seawater was successfully achieved in the three-step process using calix[4]arene derivatives as the extraction reagents in a droplet microfluidic device [51].

4.2 Precious metals

Precious metals, *i.e.* silver, gold, palladium, platinum, rhodium, iridium, osmium and ruthenium metals are indispensable metals because of their high economic value

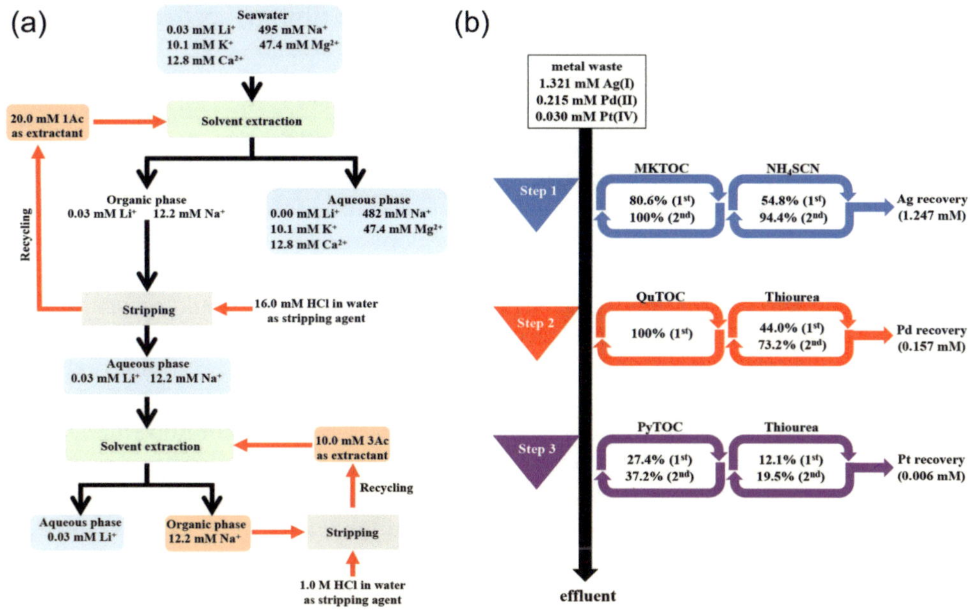

Figure 5.
(a) A process of lithium separation from seawater [51] and (b) precious metals separation from a real metal waste [52] using the droplet-based microfluidic system.

and their application for smart materials [53]. Bizzo *et al.* reported that silver(I), palladium(II) and platinum(IV) ions were found in the electronic waste serving as the secondary resource for precious metal ions [40]. Stepwise recovery of precious metals from electronic waste, therefore, must be established as an untapped resource for future precious metals supply. Even though the recovery process is complicated because of the presence of other metal ions with similar physicochemical properties, a stepwise recovery of precious metals from real metal waste using neutral type of calix[4]arene derivatives in a T-type of droplet microfluidic device was established [52]. Silver(I), palladium(II) and platinum(IV) ions extraction were carried out employing tetramethylketonic, tetraquinoyl and tetrapyridyl derivatives of calix[4]arene, respectively, as the extraction reagents. The chemical structure of tetramethylketonic, tetraquinoyl and tetrapyridyl derivatives of calix[4]arene were shown in **Figure 4** as MKTOC, QuTOC and PyTOC, respectively.

From the individual metal extraction experiment, it was found that as much as 99.9% of silver(I) was extracted at 2.00 s of extraction time from 0.10 mM of $AgNO_3$ solution in 0.10 M HNO_3 media using 5.0 mM of the tetramethylketonic derivative of calix[4]arene in a droplet microfluidic device. However, it required 72 h to achieve a similar extraction percentage of silver(I) using a batch-wise system showing that the droplet microfluidic device enhanced the required extraction time 130,000 times faster to reach the equilibrium stage of silver(I) ion extraction compared with the conventional batch-wise system. On the other hand, as much as 96.2% of palladium(II) ions was extracted at 4.00 s extraction time from 0.10 mM of $Pd(NO_3)_2$ solution in 0.10 M HNO_3 media using a tetraquinolyl derivative of calix[4]arene, and as much as 93.1% of platinum(IV) ions were extracted at 4.00 s extraction time from 0.10 mM of H_2PtCl_6 solution in 0.10 M HCl media using a tetrapyridyl derivative of calix[4]arene. The stripping of silver(I), palladium(II) and platinum(IV) ions were obtained in medium percentages using 2.00 M ammonium thiocyanate in water, 1.00 M thiourea in 1.00 M HCl and 1.00 M thiourea in water, respectively, from the metal-laden organic phase within 2.00 s only.

The evaluated real metal waste, mainly consisted of silver(I), palladium(II), platinum(IV), iron(III), copper(II), and nickel(II) ions at 1.32, 0.22, 0.03, 0.01, 0.82, and 0.01 M, respectively. To prevent any metal precipitations, the real metal waste was diluted 1000 times using 1.0 M HNO_3. Since the concentration of silver(I) was higher than palladium(II) and platinum(IV) ions, the silver(I) ions were extracted first and followed by palladium(II) and platinum(IV) ion extraction. The scheme of stepwise recovery of those precious metal ions from real metal waste is shown in **Figure 5(b)**. All extractions and stripping processes using a droplet microfluidic device were conducted at 4.00 s extraction time. Silver(I) ion extraction was carried out in two-cycle process due to high existed concentration. On the first cycle, 80.6% of silver(I) ions (1.064 mM) were extracted using 20.0 mM of the tetramethylketonic derivative of calix[4]arene. A complete extraction of silver(I) ions were achieved in the second cycle. Even though only 54.8% of silver(I) ions were stripped from the first cycle using 2.0 M ammonium thiocyanate, as much as 94.4% of silver(I) ions were recovered on the second cycle.

A complete palladium(II) ions recovery was achieved within a single cycle extraction process using 5.00 mM of the tetraquinolyl derivative of calix[4]arene, while the stripping percentages of palladium ions were 44.0 and 73.2% on the first and second cycle of stripping process using 1.00 M thiourea in 1.00 M HCl. On the other hand, 27.4 and 37.2% of platinum(IV) ions were extracted in the first and second cycles of extraction process using a tetrapyridyl derivative of calix[4]arene. Meanwhile, low stripping percentages of platinum(IV) ions, 12.1 and 19.5% of the first and second cycles of the stripping process, respectively, were obtained using 1.00 M thiourea in the water as stripping reagent. Even though platinum(IV) recovery still needs to be optimized, a stepwise and complete recovery of silver(I) and palladium(II) ions from real waste was established using a droplet microfluidic system at 4.00 s extraction time [52].

4.3 Rare-earth metals

Rare-earth metal separations leave a serious challenge due to astounding properties as well as their extensive applications in electronic and advanced materials [1]. Rare-earth metals including lanthanide and actinide groups which are classified as 4f and 5f elements in the Periodic Table. Their physicochemical properties are quite similar to each other, thus challenging many researchers to find the optimum separation process for hydrometallurgy field [30, 54]. In this part, lanthanum(III), praseodymium(III), neodymium(III), samarium(III) and uranium(VI) extraction and separation processes are described and discussed.

Lanthanum(III) extraction from a chloride solution was studied in a Y-type microfluidic device by Yin et al. By using phosphonic acid (2-ethylhexyl)-mono(2-ethylhexyl) ester (P507) as an extractant, high extraction percentage of lanthanum(III) ions was achieved within only 0.37 s [55]. Furthermore, the separation of lanthanum(III) and europium(III) ions using a T-type of droplet microfluidic device has been carried out. The separation factor of lanthanum(III) over europium(III) is in the range of 1.50 to 3.50 depends on the flow rate of the liquids [56].

Separation of praseodymium(III) and cerium(III) in a Y-type microfluidic device employing P507 as the extraction reagent was also investigated. At 12.0 s extraction time, as much as 75.0% of praseodymium(III) and 42.9% of cerium(III) ions were extracted into the organic phase yielding 3.25 as the separation factor. Even though the separation was not satisfied enough, the process could be recycled further to obtain pure cerium(III) ions as the remained aqueous phase [57].

On the other hand, neodymium(III) separation over praseodymium(III) was studied by He et al. using a Y-type droplet microfluidic device [58]. One and a half molar of P507 in sulfonated kerosene was used as the organic phase while praseodymium(III) and neodymium(III) in HCl and lactic acid media were used as the aqueous phase. At 12.0 s extraction time, it was found that P507 extracted 88.0% of neodymium(III) and 75.0% of praseodymium(III) yielding 2.23 as the separation factor of neodymium(III) over praseodymium(III). The similar separation factor was obtained using microfluidic at 12.0 s extraction time while the batch-wise system required 560 s extraction times. These findings initiated a rapid enrichment of neodymium from the wastewater sample within less than 600 s. From the wastewater containing 30–90 ppm of neodymium(III) ions, the neodymium(III) ions were enriched 200–450 times higher which is remarkable [58].

He et al. also investigated the extraction process of samarium(III) using P507 as the extraction reagent in a Y-type of droplet microfluidic device. Within 45.0 s, as much as 75.1% of samarium(III) can be extracted compared with the batch-wise system that requires 600 s to reach a similar extraction percentage [59]. Meanwhile, using a T-type of droplet microfluidic device, samarium(III) was extracted in 82.5% within 14.8 s, shorter than Y-type device. This phenomenon was caused by shorter diffusion distance and a larger specific surface area of T-type than Y-type of droplet microfluidic device [60].

Uranium(VI) extraction process was studied in a microfluidic device using several extraction reagents, such as phosphorus reagent, ionic liquids, and so on [61–64]. Among them, tributyl phosphate (TBP) exhibited as the best extraction reagent for uranium(VI) extraction process due to stable fluid dynamics, as well as, a possibility for large scale process through numbering-up the device. Darekar and coworkers investigated the extraction of uranium(VI) in 1.0 M HNO_3 using 30% v/v TBP in dodecane as the organic phase. It was found that quantitative (100%) extraction percentage was achieved at 1.50 s extraction time, however, the separation of uranium(VI) over other metal ions has not been reported yet [65].

4.4 Heavy metals

Heavy metals, especially lead(II) ions have been used in several industrial processes over the past several years and they get released into environment causing serious effects on environment and human health. A serious control monitoring of heavy metal ion concentration at industrial wastewater effluents should be comprehensively managed. So far, a lot of heavy metal removal techniques have been evaluated for wastewater treatment, however, they were unsatisfied because of heavy metals existed in trace amounts together with huge concentrations of base metal ions [5, 6]. In this part, two examples of a selective removal process of lead(II) removal using two derivatives of calix[4]arene, *i.e.* tetraacetic acid derivative (4Ac) and tetrakis(diethylamide) derivative of calix[4]arenes (EATOC) will be discussed.

In a conventional batch-wise system, the equilibrium state of lead(II) extraction using 4Ac reached after 24 h, however, it was shortened to 6.00 s using a droplet microfluidic device. The significant enhancement was observed by shorter diffusion distance and larger specific surface area aforementioned above. The lead(II) extraction percentages were increased from 73, 89 and 100% at 8.00 s for pH 2.00, 2.25 and 2.50, respectively, however, the separation of lead(II) and other metal ions was unsatisfied. As shown in **Figure 6(a)**, the extraction percentages of iron(III), copper(II), and zinc(II) were 60, 6.9 and 7.5%, respectively. This low selectivity of lead(II) extraction over Fe(III) was caused by the ion exchange mechanism in which the acetic acid groups of calix[4]arene were deprotonated and Fe(III) ions were easily extracted due to strong electrostatic interactions [66].

Another neutral calix[4]arene host molecule, EATOC was also evaluated to obtain a selective lead(II) separation from other metal ions mixture solution.

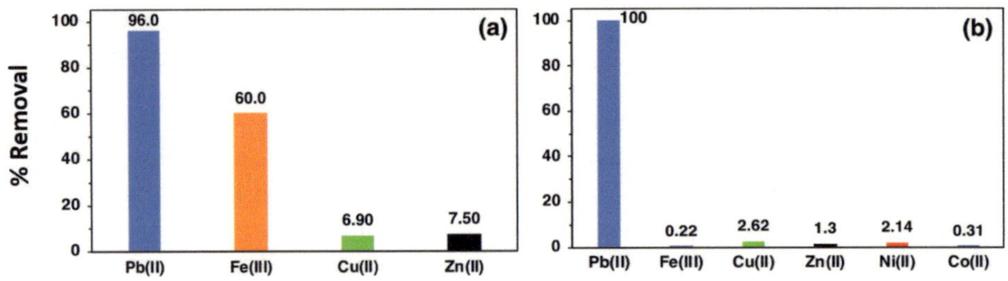

Figure 6.
Lead(II) separation from other metal ions from a competitive metal mixture solution using (a) 4Ac [66] and (b) EATOC [8] as the extraction reagents in a droplet microfluidic systems.

In this case, a neutral host molecule was used to achieve high metal selectivity. In 1.00 M HNO_3 media, one molecule tetrakis(diethylamide) derivative of calix[4]arene formed a 1:1 complex with one lead(II) ion revealed from the Job's plot experiment [8]. Using FTIR and ^{1}H-NMR titration studies, it was elucidated that the lead(II) ion was chelated by EATOC through phenoxy oxygen atoms and carbonyl oxygen moieties. Since the lead(II)-EATOC complex was stable enough, a selective lead(II) extraction (100%) was achieved from a competitive metal system containing copper(II), zinc(II), nickel(II), iron(III) and cobalt(II) ions at 0.10 mM concentration of each metal ion [8]. A complete lead(II) removal was also successfully performed from a simulated wastewater effluent containing 0.10 mM of lead(II), 45.0 mM of iron(III), while copper(II), zinc(II), nickel(II) and cobalt(II) was in 30.0 mM concentration, respectively as shown in **Figure 6(b)**. Furthermore, the lead(II) ions was easily stripped in 98% using distilled water as the stripping reagent, which is a convenient and low-price process. These finding demonstrated that droplet microfluidic system is a promising device for a complete and efficient lead(II) removal from environmental samples [8].

5. Conclusions

Several successful reports on optimized separation and recovery of alkali and base metals, precious metals, rare-earth metals, and heavy metals using the droplet microfluidic system have been published. By employing the host-guest chemistry in a droplet microfluidic system offers a rapid and efficient metal separation from binary metal mixtures or real samples either from nature or industrial activities. In order to realize the true potential of droplet microfluidic devices for renewable metallurgical applications, the improved design of microfluidics addressing larger volumes with continuous operation along with effective utilization of host-guest chemistry are greatly needed to become a step forward in building a novel metallurgical technique for selective metal separations from a real sample, which otherwise not technically feasible to extraction using the conventional batch-wise system.

Acknowledgements

This research was financially supported in part by "The Environment Research and Technology Development Fund, (No. 3K-123022) from the Ministry of Environment, Government of Japan", by JPS KAKENHI Grant Number (Grant-in-Aid for "Scientific Research (C) 16K00612) from JSPS, and by collaborative project 2019 of Institution of Ocean Energy Saga University, Japan".

Conflict of interest

The authors declare that there is no conflict of interest.

Author details

Yehezkiel Steven Kurniawan[1,2*], Ramachandra Rao Sathuluri[1,3*] and Keisuke Ohto[1]

1 Department of Chemistry and Applied Chemistry, Faculty of Science and Engineering, Saga University, Saga, Japan

2 Ma Chung Research Center for Photosynthetic Pigments, Universitas Ma Chung, Malang, Indonesia

3 Department of Reproductive Biomedicine, The National Institute of Health and Family Welfare, New Delhi, India

*Address all correspondence to: yehezkiel.steven@machung.ac.id
and sathuluri2007@gmail.com

References

[1] Binnemans K, Jones PT, Blanpain B, Gerven TV, Yang Y, Walton A, et al. Recycling of rare earths: A critical review. Journal of Cleaner Production. 2013;**51**:1-22

[2] Nishihama S, Hirai T, Komosawa I. Review of advanced liquid-liquid extraction systems for the separation of metal ions by a combination of conversion of the metal species with chemical reaction. Industrial and Engineering Chemistry Research. 2001;**40**:3085-3091

[3] Hidayah NN, Abidin SZ. The evolution of mineral processing in extraction of rare earth elements using solid-liquid extraction over liquid-liquid extraction: A review. Minerals Engineering. 2017;**112**:103-113

[4] Kaya M. Recovery of metals and nonmetals from electronic waste by physical and chemical recycling processes. Waste Management. 2016;**57**:64-90

[5] Carolin CF, Kumar PS, Saravanan A, Joshiba GJ, Naushad M. Efficient techniques for the removal of toxic heavy metals from aquatic environment: A review. Journal of Environmental Chemical Engineering. 2017;**5**:2782-2799

[6] Fu F, Wang Q. Removal of heavy metal ions from wastewaters: A review. Journal of Environmental Management. 2011;**92**:407-418

[7] Stevens GW, Perera JM, Grieser F. Metal ion extraction. Current Opinion in Colloid & Interface Science. 1997;**2**:629-634

[8] Kurniawan YS, Sathuluri RR, Iwasaki W, Morisada S, Kawakita H, Ohto K, Miyazaki M, Jumina. Microfluidic reactor for Pb(II) ion extraction and removal with amide derivative of calix[4]arene supported by spectroscopic studies. Microchemical Journal. 2018;**142**:377-384

[9] Joanicot M, Adjari A. Droplet control for microfluidics. Science. 2005;**309**:887-888

[10] Gunther A, Jensen KF. Multiphase microfluidics: From flow characteristics to chemical and material synthesis. Lab on a Chip. 2006;**6**:1487-1503

[11] Elvira KS, Solvas XC, Wootton RCR, deMello AJ. The past, present and potential for microfluidic reactor technology in chemical synthesis. Nature Chemistry. 2013;**5**:905-915

[12] Huebner A, Sharma S, Srisa-Art M, Hollfelder F, Edel JB, deMello AJ. Microdroplets: A sea of applications? Lab on a Chip. 2008;**8**:1244-1254

[13] The SY, Lin R, Hung LH, Lee AP. Droplet microfluidics. Lab on a Chip. 2008;**8**:198-220

[14] Yao X, Zhang Y, Du L, Liu J, Yao J. Review of the applications of microreactors. Renewable and Sustainable Energy Reviews. 2015;**47**:519-539

[15] Ciceri D, Perera JM, Stevens GW. The use of microfluidic devices in solvent extraction. Journal of Chemical Technology and Biotechnology. 2013;**89**:771-786

[16] Wang K, Luo G. Microflow extraction: A review of recent development. Chemical Engineering Science. 2017;**169**:18-33

[17] Mary P, Studer V, Tabeling P. Microfluidic droplet-based liquid-liquid extraction. Analytical Chemistry. 2008;**80**:2680-2687

[18] Mason LR, Ciceri D, Harvie DJE, Perera JM, Stevens GW. Modelling of interfacial mass transfer in microfluidic solvent extraction part I. Heterogenous transport. Microfluidics and Nanofluidics. 2013;**14**:197-212

[19] Ciceri D, Mason LR, Harvie DJE, Perera JM, Stevens GW. Modelling of interfacial mass transfer in microfluidic solvent extraction: Part II. Heterogenous transport with chemical reaction. Microfluidics and Nanofluidics. 2013;**14**:213-224

[20] Sahu A, Vir AB, Molleti LNS, Ramji S, Pushpavanam S. Comparison of liquid-liquid extraction in batch systems and micro-channels. Chemical Engineering and Processing. 2016;**104**:190-200

[21] Lehn JM. Supramolecular chemistry: Where from? Where to? Chemical Society Reviews. 2017;**46**:2378-2379

[22] Kollman PA. The nature of molecular recognition: Examples from host/guest chemistry. In: Wipff G, editor. Computational Approaches in Supramolecular Chemistry. NATO ASI Series (Series C: Mathematical and Physical Sciences). Dordrecht: Springer; Vol. 426. 1994. pp. 175-181

[23] Cram DJ, Cram JM. Host-guest chemistry. Science. 1974;**183**:803-809

[24] Kim HJ, Lee MH, Mutihac L, Vicens J, Kim JS. Host-guest sensing by calixarenes on the surfaces. Chemical Society Reviews. 2012;**41**:1173-1190

[25] Gutsche CD, Dhawan B, No KH, Muthukrishnan R. Calixarenes. 4. The synthesis, characterization, and properties of the calixarenes from p-tert-butylphenol. Journal of the American Chemical Society. 1981;**103**:3782-3792

[26] Ohto K. Review of the extraction behavior of metal ions with calixarene derivatives. Solvent Extraction Research and Development, Japan. 2010;**17**:1-18

[27] Ludwig R. Calixarenes in analytical and separation chemistry. Fresenius' Journal of Analytical Chemistry. 2000;**367**:103-128

[28] Silwa W, Girek T. Calixarene complexes wih metal ions. Journal of Inclusion Phenomena and Macrocyclic Chemistry. 2010;**66**:15-41

[29] Rebily JN, Colasson B, Bistri O, Over D, Reinaud O. Biomimetic cavity-based metal complexes. Chemical Society Reviews. 2015;**4**:467-489

[30] Atanassova M, Kurteva V. Synergism as a phenomenon in solvent ectraction of 4f-elements with calixarenes. RSC Advances. 2016;**6**:11303-11324

[31] Ludwig R, Dzung NTK. Calixarene-based molecules for cation recognition. Sensors. 2002;**2**:397-416

[32] Hodgson GW, Charles ME. The pipeline flow of capsules: Part 1: The concept of capsule pipelining. Canadian Journal of Chemical Engineering. 1963;**41**:43-45

[33] Burns JR, Ramshaw C. The intensification of rapid reactions in multiphase systems using slug flow in capillaries. Lab on a Chip. 2001;**1**:10-15

[34] Kurniawan YS. Micro total analysis system application for biomedicals: A mini-review. Biomedical Journal of Scientific & Technical Research. 2019;**12**(4):1-2

[35] Kurniawan YS, Imawan AC, Sathuluri RR, Ohto K, Iwasaki W, Miyazaki M, et al. Microfluidics era in chemistry field: A review. Journal of the Indonesian Chemical Society. 2019;**2**(1):7-23

[36] Kaminski TS, Garstecki P. Controlled droplet microfluidic system for multistep

chemical and biological assays. Chemical Society Reviews. 2017;**46**:6210-6226

[37] Kashid MN, Agar DW, Turek S. CFD modelling of mass transfer with and without chemical reactions in the liquid-liquid slug flow microreactor. Chemical Engineering Science. 2007;**62**:5102-5109

[38] Ohto K, Kim JY, Morisada S, Maeki M, Yamashita K, Miyazaki M. Microreactor extraction system with macrocyclic host compounds for rare metal recovery. International Journal of the Society of Materials Engineering for Resources. 2014;**20**:92-96

[39] Ghaini A, Kashid MN, Agar DW. Effective interfacial area for mass transfer in the liquid-liquid slug flow capillary microreactors. Chemical Engineering and Processing. 2010;**49**:358-366

[40] Bizzo WA, Figueiredo RA, de Andrade VF. Characterization of printed circuit boards for metal and energy recovery after milling and mechanical separation. Materials. 2014;**7**:4555-4566

[41] Yang L, Zhao Y, Su Y, Chen G. An experimental study of copper extraction characteristics in a T-junction microchannel. Chemical Engineering and Technology. 2013;**36**:985-992

[42] Jiang F, Pei J, Yin S, Zhang L, Peng J, Ju S, et al. Solvent extraction and stripping of copper in a Y-Y type of microchannel reactor. Minerals Engineering. 2018;**127**:296-304

[43] Ciceri D, Mason LR, Harvie DJE, Perera JM, Stevens GW. Extraction kinetics of Fe(III) by di-(2-ethylhexyl) phosphoric acid using a Y-Y shaped microfluidic device. Chemical Engineering Research and Design. 2014;**92**:571-580

[44] Le T, Xiao B, Ju S, Peng J, Jiang F. Separation of indium from impurities in T-type microreactor with D2EHPA. Hydrometallurgy. 2019;**183**:79-86

[45] Jahromi PF, Karimi-Sabet J, Amini Y. Ion-pair extraction-reaction of calcium using Y-shaped microfluidic junctions: An optimized separation approach. Chemical Engineering Journal. 2018;**334**:2603-2615

[46] Marsousi S, Karimi-Sabet J, Moosavian MA, Amini Y. Liquid-liquid extraction of calcium using ionic liquids in spiral microfluidics. Chemical Engineering Journal. 2019;**356**:492-505

[47] Tamagawa O, Muto A. Development of cesium ion extraction process using a slug flow microreactor. Chemical Engineering Journal. 2011;**167**:700-704

[48] Swain B. Recovery and recycling of lithium: A review. Separation and Purification Technology. 2017;**172**:388-403

[49] Muto A, Hirayama Y, Tokumoto H, Matsuoka A, Noishiki K. Liquid-liquid extraction of lithium ions using a slug flow microreactor: Effect of extraction reagent and microtube material. Solvent Extraction and Ion Exchange. 2016;**35**:61-73

[50] Hirayama Y, Hinoue M, Tokumoto H, Matsuoka A, Noishiki K, Muto A. Liquid-liquid extraction and separation of cobalt and lithium ions using a slug flow microreactor. Journal of Chemical Engineering of Japan. 2018;**51**(3):222-228

[51] Kurniawan YS, Sathuluri RR, Ohto K, Iwasaki W, Kawakita H, Morisada S, et al. A rapid and efficient lithium-ion recovery from seawater with tripropyl-monoacetic acid calix[4]arene derivative employing droplet-based microfluidic reactor system. Separation and Purification Technology. 2019;**211**:925-934

[52] Sathuluri RR, Kurniawan YS, Kim JY, Maeki M, Iwasaki W,

Morisada S, et al. Droplet-based microreactor system for stepwise recovery of precious metal ions from real metal waste with calix[4]arene derivatives. Separation Science and Technology. 2018;**53**(8):1261-1272

[53] Ding Y, Zhang S, Liu B, Zheng H, Chang CC, Ekberg C. Recovery of precious metals from electronic waste and spent catalysts: A review. Resources, Conservation and Recycling. 2019;**141**:284-298

[54] Chen Z, Wang WT, Sang FN, Xu JH, Luo GS, Wang YD. Fast extraction and enrichment of rare earth elements from waste water via microfluidic-based hollow droplet. Separation and Purification Technology. 2017;**174**:352-361

[55] Yin SH, Zhang LB, Peng JH, Ju SH, Zhang LH. Microfluidic solvent extraction of La(III) with 2-ethylhexyl phosphonic acid-2-ethylhexyl ester (P507) by a microreactor. Chemical Engineering and Processing. 2015;**91**:1-6

[56] Zhang H, Wang H, Luo X, Leung DYC, Pang Q, Xu H, et al. Toward a mechanistic understanding of microfluidic droplet-based extraction and separation of lanthanides. Chemical Engineering Journal. 2019;**365**:673-679

[57] Yin S, Chen K, Srinivasakannan C, Li S, Zhou J, Peng J, et al. Microfluidic solvent extraction of Ce (III) and Pr (III) from a chloride solution using EHEHPA (P507) in a serpentine microreactor. Hydrometallurgy. 2018;**175**:266-272

[58] He Y, Chen K, Srinivasakannan C, Li S, Yin S, Peng J, et al. Intensified extraction and separation Pr (III)/Nd (III) from chloride solution in presence of a complexing agent using a serpentine microreactor. Chemical Engineering Journal. 2018;**354**:1068-1074

[59] He Y, Pei J, Srinivasakannan C, Li S, Peng J, Guo S, et al. Extraction of samarium using a serpentine Y-junction microreactor with 2-ethylhexyl phosphonic acid mono-2-ethylhexyl. Hydrometallurgy. 2018;**179**:175-180

[60] He Y, Zeng Y, Chen K, Ullah E, Li S, Zhang L, et al. Solvent extraction performance of Sm(III) using a T-junction microreactor with 2-ethylhexyl phosphonic acid mono-2-ethylhexyl (EHEHPA). Chemical Engineering and Processing. 2019;**136**:28-35

[61] Darekar M, Singh KK, Mukhopadhyay S, Shenoy KT, Ghosh SK. Solvent extraction in microbore tubes with UNPS-TBP in dodecane system. Separation and Purification Technology. 2014;**128**:96-106

[62] Tsaoulidis D, Dore V, Angeli P, Plechkova NV, Seddon KR. Dioxouranium(VI) extraction in microchannels using ionic liquids. Chemical Engineering Journal. 2013;**227**:151-157

[63] Zhang J, Liao JS, Deng DC, Wang HS, Zhen ZL. Solvent extraction of uranium by laminar flow in microfluidic chips. Advances in Materials Research. 2013;**709**:485-490

[64] Tsaoulidis D, Dore V, Angeli P, Plechkova NV, Seddon KR. Extraction of dioxouranium(VI) in small channels using ionic liquids. Chemical Engineering Research and Design. 2013;**91**(4):681-687

[65] Darekar M, Singh KK, Sapkale P, Goswami AK, Mukhopadhyay S, Shenoy KT. On microfluidic solvent extraction of uranium. Chemical Engineering and Processing. 2018;**132**:65-74

[66] Kurniawan YS, Ryu M, Sathuluri RR, Iwasaki W, Morisada S, Kawakita H, et al. Separation of Pb(II) ion with tetraacetic acid derivative of calix[4]arene by using droplet-based microreactor system. Indonesian Journal of Chemistry. 2019;**19**(2):368-375

Chapter 2

Advances in Droplet Microfluidics with Off-the-Shelf Devices and Other Novel Designs

Maxine Yew, Kaiseng Koh and Yong Ren

Abstract

For the past three decades since the dawn of the concept of scaling down fluidic processes and systems, various microfabrication techniques have been significantly explored and developed. Glass and elastomers, as first-generation microfluidic device materials, are still widely used, while other alternatives have emerged. We have seen a rise in novel and innovative device designs and fabrication in droplet microfluidics which enable easier and more cost-effective approach of generating droplets, such as the use of commercially available "off-the-shelf" components, plug-and-play modular devices, and 3D-printed droplet generators. This chapter aims to review some of these facile approaches for droplet generation and discuss the versatility and functionability of these designs for possible commercial formulations. Discussions of rationales for and challenges of the development of these conceptual devices are included.

Keywords: off-the-shelf, 3D printing, droplet generator, modular, parallelization

1. Introduction

Microfluidics defines the science and engineering of a small scale fluid system. It is a study of design, fabrication, and operation of a system conducting fluids in microscopic channels of widths or diameters ranging from 10 to 500 micrometers (μm). Gañán-Calvo first introduced the use of micron-scale capillaries later established by Thorsen et al. who demonstrated the use of a simple T-junction microfluidic device to control the flow of immiscible liquids, forming monodisperse droplets [1, 2]. The manipulation of discrete fluid packets in microdroplets provides benefits in large reduction of reagent volume, size of sample, and the equipment [3]. This opened a completely new wave of interest in microfluidic systems in a broad range of fields such as biochemical, engineering, and pharmaceutical benefits from the advantages of small and precise.

Droplet-based microfluidics, one subcategory of microfluidics, focuses on the discrete volume creation with the use of immiscible fluids and allows the handling of fluids under confined spatial and temporal control. Applications of droplet microfluidics are largely distinguished into two aspects: droplet reactors for biochemical reactions and analysis [4] and droplet-templated material synthesis. The proverbial use of lab-on-a-chip (LOC) in various biochemical analyses is driven by the time and cost-efficiency of performing automated high-throughput analysis with small volume of reagents, especially in the areas of genomic study, point-of-care (POC) diagnostic, drug discovery, etc. To date, only a handful of selected microchips have been commercialized [5]. Compared to conventional bulk methods, microfluidic techniques provide a power platform that enables the creation of highly controllable emulsion droplets in which the size, shape, and composition of the droplet can be manipulated. This allows the synthesis of functional droplets such as microcapsules for drug delivery and cell encapsulation, microparticles, and various types of Janus particles.

With the rapid development in microfluidic systems and its applications, different selections of techniques and materials are increasingly accessible for fabrication of microsystems (especially for those who are just starting out in the field). Some of these fabrication techniques require high-priced tools and machineries with skilled workers to produce intricate designs of microchannels and microchambers in which droplets are generated and manipulated, while simple assemblies of microdevices with commercially available components have also been reported. To produce emulsions, which are droplets in immiscible phase, the selection of materials with compatible surface wettability is important.

The huge potential of droplet-based microfluidics has stimulated rapid development of flexible platforms to generate highly monodispersed droplets with diverse structures. The recent years see a soaring number of researches and publications on life sciences and material synthesis applications based on microfluidic approach. Likewise, there has also been a rise in novel and innovative device designs and fabrication in droplet microfluidics which supposedly enable easier and more cost-effective approach of generating droplets, such as the use of commercially available "off-the-shelf" components, plug-and-play modular devices, and 3D-printed droplet generators. While this signifies a positive outlook on the development of microfluidics, the ultimate reception of the technology depends primarily on the end users. Different applications where microfluidic approach is employed have different prerequisites for commercialization, for instance, for material synthesis such as catalyst preparation or drug formulations, high throughput is expected to generate sufficient materials, whereas for chemical/biological assays, throughput might not be a focal criterion, instead development of chip-to-world interfaces for stable connection to analytical units would be crucial. Up until now, the successfully commercialized microfluidic products are mainly for POC diagnostics, genotyping and sequencing, and other biomedical-related applications [5].

2. Summary of existing droplet-based microfluidic technology

As the standard material interface used in the field of microelectronics which pioneered microfluidics, glass and silicon are naturally the first-generation materials for microfluidic devices [6]. Originally, glass and silicon substrates are processed via standard photolithography which mainly comprises three steps: (i) lithography or pattern transfer onto a substrate, (ii) etching of microchannels, and (iii) lastly bonding to create an enclosed structure [6]. The process of photolithography has been well documented. Devices fabricated via lithographic methods generally have quasi-two-dimensional (2D) planar flow [7]. Umbanhowar et al. and Utada et al. both introduced three-dimensional (3D) co-flow and hydrodynamic flow-focusing devices with tapered capillaries to produce monodisperse single and multiple emulsions, respectively [8, 9]. Both dispersed and continuous phase fluids meet in parallel streams in the co-flow geometry, while for the flow-focusing device, the dispersed and continuous phases are introduced in opposite directions, whereby the flow is eventually focused through an orifice. By combining both geometries, monodisperse multi-emulsions could be generated via single-step emulsification (see **Figure 1**). 3D axisymmetric flow-focusing devices circumvent the wetting of channel walls by the dispersed phase as the effect of wall channel is of less concern and droplet formation frequency is higher [10–12]. Microgrooves can also be formed on glass substrates via deep reactive ion etching (DRIE). Nisisako and Torii have produced a planar synthesis silica glass chip with 256 parallel generators formed by DRIE which could produce emulsions up to a rate of 320 mL/h [13, 14]. However, micromachining of glass is very expensive, and it gets even more difficult and costly to fabricate elaborated and complex designs on a glass chip.

Soft lithography, an extension from conventional photolithography, allows the lithographic master to be used for rapid prototyping of elastomeric material such as polydimethylsiloxane (PDMS). PDMS has been and is still a popular option because it is cheap and elastic and it is easy to work with high flexibility [15]. Its intrinsic properties such as high permeability to gases and optical transparency have even greatly extended its use in biomedical researches such as cell culturing and tissue analysis [16]. PDMS devices (see **Figure 2**) manufactured via soft lithography are inherently 2D; however, fabrication of 3D microfluidic devices is now possible with advances in 3D printing, whereby 3D PDMS microchannel can be casted from 3D-printed molds [17].

With the vast selections available, researchers often establish their choices for the development of microfluidic systems based on the applications, material compatibility, ease of fabrication, as well as the start-up cost incurred; many methods/techniques remain underdeveloped or ceased to be developed presumably due to high start-up cost. While glass and PDMS have been extensively used in pioneering many microfluidic researches and applications, the known disadvantages associated with the fabrication and use of these materials have often been deliberately overlooked by the microfluidic community. As PDMS is inherently hydrophobic, small hydrophobic nonpolar molecules tend to be absorbed into the polymer, which could significantly disrupt the microenvironment at which biological studies are carried out [15]. Economy-wise, it would be very costly to work with glass due to laborious procedures and difficulty in reproducing, while expansion of PDMS for industrial work will be less viable. Apart from glass and PDMS, plastic has also been a popular option and one that is believed to play a major role in translating microfluidic research into commercialized technologies [19]. There has been an increasing interest in developing materials such as polymethymethacrylate (PMMA), polycarbonate, and cyclic olefin (co)polymers through microfabrication techniques such as hot embossing, injection molding, micromilling, and stereolithography [20, 21]. Plastics are more rigid and hence more collapse resistant than PDMS, and they generally have good compatibility for biological applications [19].

Hot embossing and injection molding are both replicating techniques akin to soft lithography, with the use of thermoplastic materials. With hot embossing, a thermoplastic film is patterned against a master, and a cast is formed as the molds

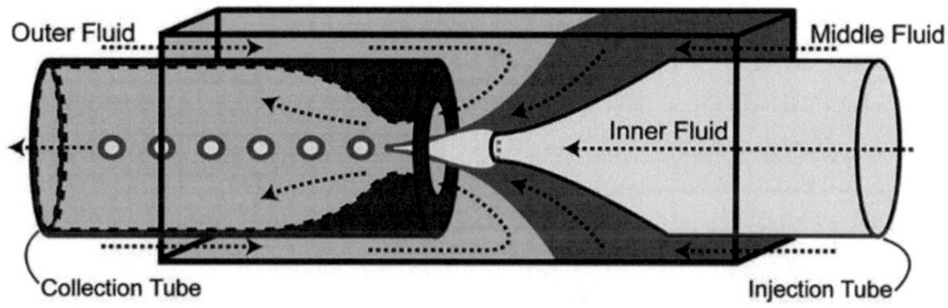

Figure 1.
Glass capillary microfluidic device for double emulsion formation through coaxial jet [9].

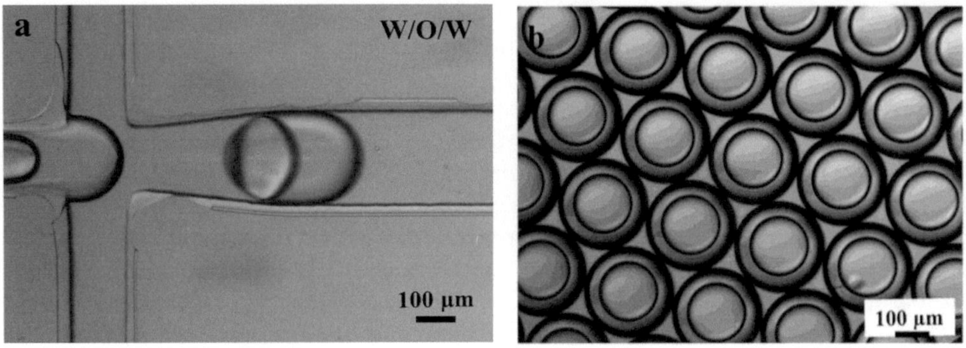

Figure 2.
Double emulsions produced from a PDMS device replicated via soft lithography. Part of the device is subjected to plasma treatment to render it hydrophilic. (a) Optical image of a double emulsion forming at the second junction of device. (b) Double emulsions collected with high uniformity. Reproduced from Ref. [18] with permission from the Royal Society of Chemistry.

are heated and compressed [22]. Injection molding involves the heating and subsequent injection of a thermoplastic melt into mold cavity, after which the cast will be removed from the mold once cooled down. While the resolution of molds and casts for both methods is subject to the methods used to make the mold inserts, especially complex three-dimensional structures, such replicating techniques are promising for rapid prototyping of microdevices. Micromilling, which has been a less popular alternative compared to mold replicating, is a conventional machining approach of incising features on workpieces. Advances in the milling technology now allow 3D computer-aided design (CAD) models to be prototyped on automation through computer numerical control (CNC) mills with microscale resolution [19]. Micromilling can also be used to fabricate molds to be used in replicating devices.

While these microfabrication techniques are more established now than two decades ago, ongoing researches are directed at reducing the cost of the prototyping techniques with cheaper materials. Currently, hot embossing and injection molding are limited to thermoplastics only, while the main concern for micromilling would be whether the quality of micromilled devices would suffice for practical use.

3. Some novel and innovative designs of droplet generators

3.1 Off-the-shelf devices

With rapid development in droplet emulsification, there has been much consideration in the use of "off-the-shelf" devices. While the term "off-the-shelf microdevice" might be ambiguous and may refer to already commercialized microchips, here it is defined as droplet generators assembled using commercially available components. Some of these components include medical dispensing needles, laboratory tubing, and tiny plastic fittings (such as ferrules and flanges) that might have been used as universal fluidic fittings or connectors to, for instance, laboratory analytical units. These easily assembled 3D devices are seen as cheap yet practical alternatives to conventional droplet generators requiring laborious effort to be fabricated. Simply said there could be no need for lithographical pattern transfers, microforging of fine glass capillaries, or any other hefty microfabrication units for the synthesis of these devices. Such robust and disposable devices from quick assembly of commercially available components are also useful and cost-saving when hazardous experiments are conducted [23]. Unsurprisingly, there have been some publications that explore and incorporate the use of such droplet generators for production of functional materials such as microcapsules for drug delivery or gas sorption [24, 25]. Apart from being highly cost-effective and easy-to-make, some have reported that their off-the-shelf devices could also be easily disassembled for cleaning and there is also higher flexibility in droplet generation as the device can be reconfigured for desired formation [26].

Figure 3 shows the different devices assembled with components from commercial microfluidic sources, IDEX Health & Science. In **Figure 3a**, a 3D microfluidic nozzle is formed using a micrometering valve assembly and other fluidic fittings in an effort to focus and inject particulate samples into a downstream fused silica microfluidic device [27]. A silica capillary with a cone shape tip is used as the delivering nozzle and is sealed to the microferrule with a 5-min epoxy. **Figure 3e** and **f** shows a capillary microfluidic device with two polyetheretherketone (PEEK) chromatography tees, while **Figure 3d** shows a micro-cross droplet generator with high-pressure PEEK adapters [26, 28]. Microferrules and female nuts are used to secure the capillaries in place. According to Benson et al., the device can produce both oil-in-water (O/W) and water-in-oil (W/O) emulsions by just flushing with ethanol before the formation of the other, and the device is also reusable and could be disassembled for cleaning. The tips of two glass capillaries are flamed to create two orifices: one for the injection of the inner fluid and the other for the focusing of the flow. With the flamed-tip design, there is no need to modify the wettability of the capillary, allowing changeable formation of O/W and W/O. The distance between the orifices of the two aligned capillaries can also be adjusted, and the flow regime in the device changes even when the flow rates are maintained. Wu et al. compared the synthesis of droplets in the micro-cross droplet generator to that from a planar PDMS chip having a similar geometry. The former has a higher droplet formation frequency due to faster 3D fluid thread collapse than in the 2D flow. The micro-cross droplet generator is also able to operate under high flow pressure of up to 400 psi [28].

Steinbacher et al. demonstrated an assembly of a simplified mesofluidic device entirely from off-the-shelf components, small-diameter tubing, barbed tubing adapters, and needles, into droplet-forming devices and particle concentrator [23]. The readily available components can be arranged into a T-junction

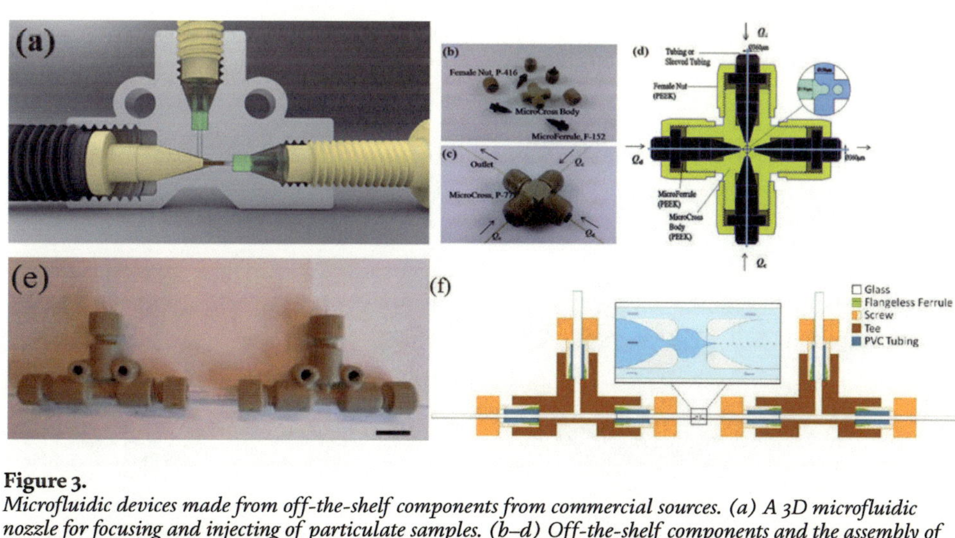

Figure 3.
Microfluidic devices made from off-the-shelf components from commercial sources. (a) A 3D microfluidic nozzle for focusing and injecting of particulate samples. (b–d) Off-the-shelf components and the assembly of a micro-cross droplet generator. (e and f) Image and schematic of a reusable modular glass capillary device assembled from commercial materials. Scale bar = 1 cm. Reproduced from Refs. [26-28] with permission from the Royal Society of Chemistry.

device or a flow-focusing configuration, and the formation of monodisperse droplets, microcapsules, and Janus particles has been reported to be successful, with the results comparable to those synthesized with more complicated devices. Li et al. also reported the assembly of yet another entirely off-the-shelf micro-device with flexible designs [29]. The device is mainly composed of dispensing needles of different sizes (60–1550 μm) and assembled with mini tee- and cross-links as shown in **Figure 4**. The stainless-steel dispensing needles are arranged and aligned to form microchannels for the flow of the immiscible phases. The assembling of the device is made using plastic tubing as well as UV curable glue. Monodisperse single and multiple droplets have been successfully formed using the needle-based microdevice by reconfiguring the device. The size of the droplets can be controlled by varying the needle used as well as the flow rates of each phase.

The past few decades have seen a myriad of microfluidic device designs, yet the innovative assemblies of off-the-shelf devices have shown that the basic modalities of a droplet generator could be substituted by commercially available components, tee-link for a T-junction, cross-link for a cross-channel, dispensing needles and glass capillaries as the inlet and flow channels, and other structures that could also be replaced by commercially available parts, which greatly reduce the fabrication cost to as low as 3USD per device. Assemblies of such devices have also shown that without sealing the droplet generators onto rigid substrates, the devices are recyclable and can be reconfigured for versatile droplet generation. The development of such microdevices has seemingly negated the need for complicated microfabrication techniques such as those listed previously. However, the need for manual assembly as well as the difficulty in aligning glass capillaries remains an issue in the fabrication of these devices as most of them still adopt the use of glass capillaries, especially when nontransparent fittings are used, it poses an obstacle for clear observation of droplet formation. Alas, there has not been many further researches reported following successful demonstrations of the use of these devices, and it is questionable if these devices would stand any chance to be commercialized. To the best of the author's knowledge, there has yet to be any work reported on the parallelization of off-the-shelf devices. In the meantime, such devices remain a short-term solution for preliminary droplet generation studies and a conceptual proving tool for new droplet-based applications. Nonetheless they would be useful for applications in detection or quick analyses, given their versatility to be connected or integrated to other existing units.

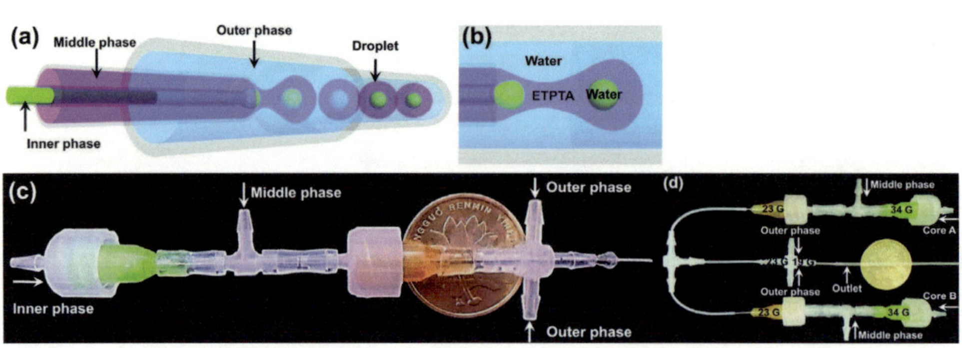

Figure 4.
(a) Schematic illustration of generation of double emulsions from needle-based coaxial microfluidic device. (b) Clear depiction of production of W/O/W emulsions. (c) The assembly of the needle-based device with a coin for scale. (d) Extended configuration of needle-based microdevice for complex emulsion production. Reproduced from Ref. [29] with permission from the Royal Society of Chemistry.

3.2 3D-printed devices

Additive manufacturing (3D printing) has been vastly adopted in fabrication of microfluidic devices with the advance of the technology in generating structures at increasingly high precision. 3D printing is a layer-by-layer manufacturing technology that now allows an extensive range of materials to be printed, such as various types of plastics and even glass [20, 30]. As is it largely automated and assembly-free, rapid prototyping can be achieved for microdevices. There are different types of 3D printing techniques, but they are primarily categorized into (i) extrusion-based 3D printing, (ii) stereolithography, and (iii) multijet (or polyjet) modeling printing [17]. 3D printing of microfluidic devices with elaborated and complex designs is now possible with the advances in the resolution and accessibility of 3D printers. In this section, an emphasis is given to 3D-printed modular devices, to introduce and highlight the versatility of droplet generators with "plug-and-play" functionality, which could be a good solution for chip-to-world integration and development.

Vijayan and Hashimoto developed 3D-printed fittings to form axisymmetric flow-focusing droplet generators with other readily available materials such as needles and elastic tubes [31]. While monolithic 3D-printed device is now possible with the rapid development in 3D printing technology, the authors have opted to print the device in parts and assemble with other commercially available parts, as they claim that this gives more customizability to the device's configuration, allowing for the generation of higher order and complex emulsions. Akin to the off-the-shelf needle device by Li et al., commercially available dispensing needles of varying inner diameters (60–250 μm) are used to produce droplets of a larger range of sizes, while for fully 3D-printed microchannels, the minimum channel features achievable are a few hundred micrometers [32]. Multiple emulsions and Janus particles can also be formed by connecting the device in series or incorporating a 3D-printed branched Y-channel such as that shown in **Figure 5g**.

For stereolithography and polyjet 3D printing, post-processing is needed to remove support materials or nonpolymerized resins from narrow channels; and it is even more challenging and time-consuming to do so for monolithically printed devices [31, 32]. Modularly printed microfluidic devices, however, allow easy removal of uncured materials and surface treatment onto different modules to facilitate generation of complex emulsions, and external active components may also be incorporated to the device as individual modules. Such modular plug-and-play units would also be useful in chip-to-world interface to existing systems requiring microfluidic solutions. Many have produced comprehensive reviews on the advances of 3D-printed devices, including works on plug-and-play modular devices, and the works included in the following discussion merely demonstrate the potential of 3D printing in microfluidics and hence are far from exhaustive.

Ji et al. presented a modular multimaterial 3D-printed device that allows both passive and active (pneumatic control) generations of various emulsions [11]. Three modules have been printed, a function module, a T-junction module, and a co-flow module, of which at least two modules are needed for the assembly of a functional device. There are three switchable modes for the function modules, namely, single-inlet and dual-inlet modules, depending on the type of emulsions to be generated or a pneumatic control unit (PCU), for active control of droplet generation by controlling the deformation of the flexible channel. A multimaterial polyjet 3D printer is used to print the PCU module which consists of both elastic (orange) and rigid (gray) parts, as indicated in **Figure 6a**, while the T-junction and co-flow modules are printed with a stereolithography printer using a clear resin. Surface treatment is carried out on individual modules to render the surface hydrophobic/hydrophilic. **Figure 6d–f** shows passive generation of single and multiple emulsions in the differently configured device. Likewise, single and double emulsions can be generated by

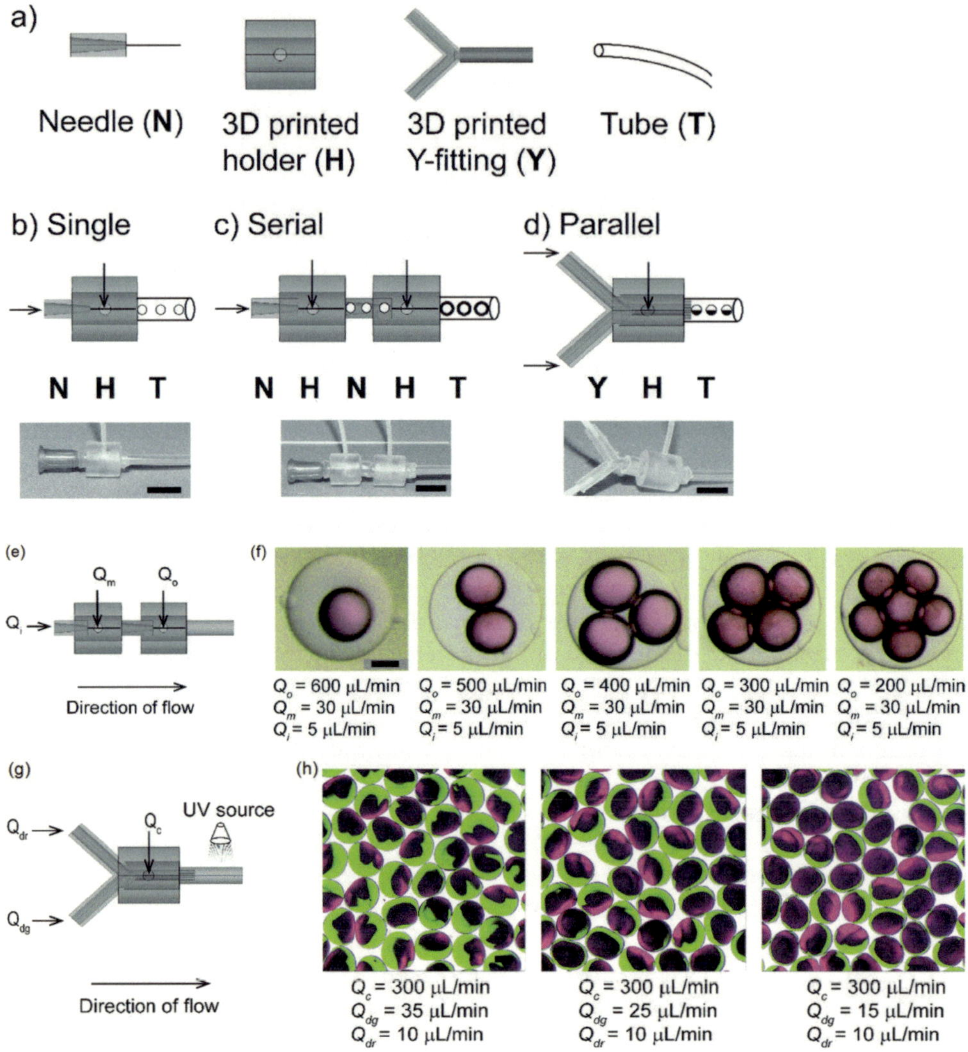

Figure 5.
(a) 3D-printed modules and fittings for the fabrication of microfluidic axisymmetric flow-focusing device: needle (N), 3D-printed holder (H), 3D-printed Y-fitting (Y), and tube (T). Schematics and optical images of (b) a single flow-focusing device, (c) serially connected flow-focusing devices to produce double emulsions, and a (d) flow-focusing device with a Y-shaped parallel laminar flow to produce compartmented particles. Scale bar = 10 mm. (e) a schematic illustration of serially connected flow-focusing devices for producing double emulsions. (f) W/O/W double emulsions with varying numbers of inner droplets. Scale bar = 600 µm. (g) Schematic illustration of production of bi-compartmented Janus particles with a Y-channel as inlet. (h) Optical micrographs of produced compartmented particles at different flow conditions. Scale bar = 300 µm [31]. Published by the Royal Society of Chemistry.

replacing the single–/dual-inlet module with a PCU, and under different applied air pressure, the flexible channel undergoes deformation through pneumatic excitation, and regimes of unstable and stable droplet generations are identified. The modules are connected through snap-fit joints printed to the modules with an O-ring, and no leakage is detected during the experiment even at a pressure of 4 bar.

Song et al. also fabricated a modular microfluidic system that allows both passive and active generation and manipulation of complex emulsion droplets [33]. Various modules such as inlet/outlet, I-junction, Y-junction, co-flow droplet generator, mixer, and a module containing electrode are printed with an i3DP printer with clear transparent UV curable polymer; however, the visibility of the modules is slightly reduced

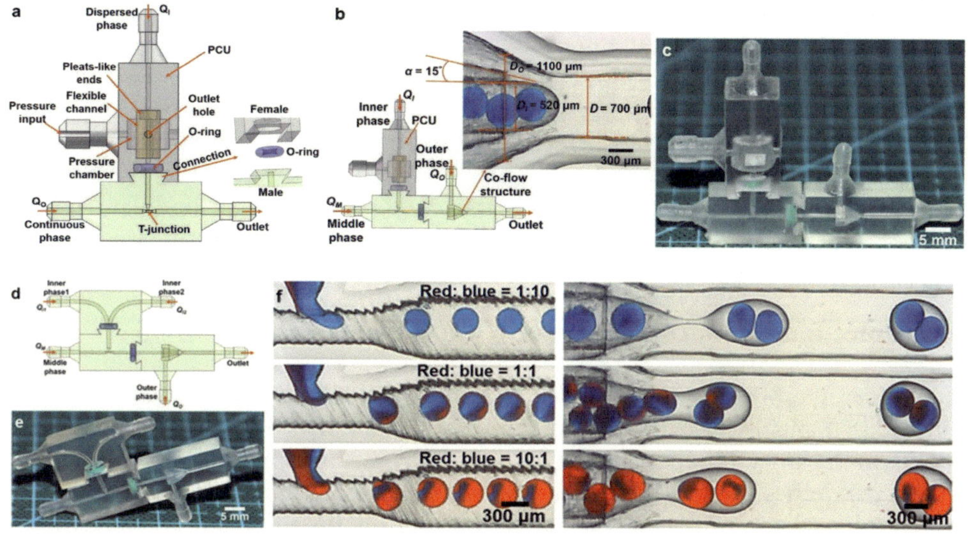

Figure 6.
3D-printed modular device. (a) Schematic of a single droplet-generating pneumatic device consisting of a PCU and a T-junction. (b and c) Schematic and optical image of a pneumatic device for generation of double or multicore emulsions. (d and e) Schematic and optical image of modular device with dual inlet. (f) Frames from high-speed recording showing complex emulsions produced via dual-inlet device [11].

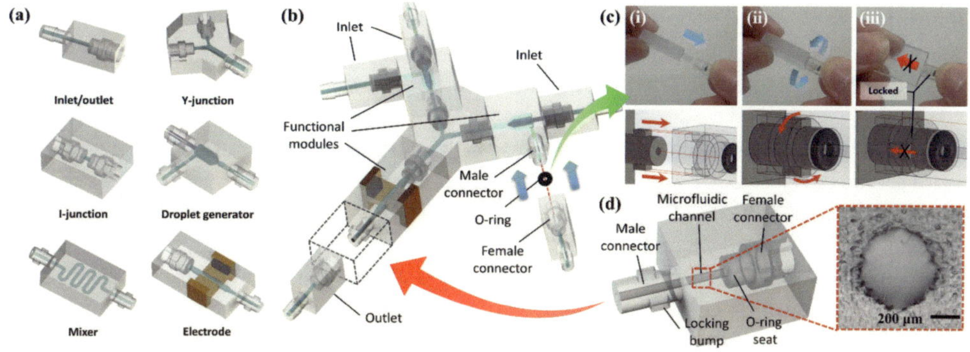

Figure 7.
Schematic of 3D-printed modular microfluidic system. (a) 3D-printed modules of different functions and structures. (b) A full assembly of the modular device. (c) The design of the interlocking male and female connectors. (d) Male and female connectors printed on each module [33].

due to surface roughness. The modules have interlocking male and female connectors that secure the assembly, and an inherently printed O-ring groove around the female connectors circumvents the leakage around the joint. An assembly of all the modules into an operative device is depicted in **Figure 7**. When the generated droplets reached the electrode module, they experience an electric field in a direction perpendicular to the flow direction. The shape of the droplet is distorted as an effect to the electro-hydrodynamic flow patterns within and outside the droplet. Again, by reconfiguring the device, versatile range of droplets has been formed such as alternating droplets and single- and dual-core double emulsions. Under the influence of the electric field, the droplets are subjected to electrocoalescence to form Janus structure.

Unlike other methods, 3D printing is an additive technique, and 3D parts are formed layer by layer without requiring molds. A variety of 3D printing methods has been developed; likewise, a wider range of choices of polymer-based printing materials are available, although some are proprietary materials limited to certain printing techniques. At present, 3D-printed devices are limited by the minimum achievable channel dimensions, subject to the printing resolution. The printing time of a device is also subject to the complexity of its design structure. Nonetheless, given the current development trajectory in additive manufacturing, there is high anticipation on 3D printing to resolve challenges with current existing microfabrication techniques such as high start-up cost, low channel resolution, and manual assembly of devices. The demonstration of plug-and-play modular devices is believed to be able to partly resolve the long-standing chip-to-world issues.

4. Droplet-based microfluidics: now and then

Droplet microfluidics is an enabling platform that is now widely recognized and adopted in many research disciplines. The past few decades have seen a rapid evolution in microfluidic-based technologies, especially in contributing to life science-related researches and gradually into material sciences. Much could be benefitted and have been realized from the subscale world of micros and nanos, attributed to rapid analyses and fast reactions, high precision, and good control of droplet size and compositions. Progressively, droplet-based platforms have been used for the synthesis of functional materials such as drug delivery vehicles and nanomaterials [34]. Conversely, the low production rate of droplets at 0.1–10 mL/h hinders the extensive use of the technology for high-volume production [35]. The direct solution to addressing low throughput of droplets is to increase the number of droplet generators to scale up production of droplets within a single chip.

4.1 Microfluidic emulsification: toward parallelization

Such effort to scale up the production of chip-based emulsions has been reported as early as 2008 by Nisisako and Torii who demonstrated parallelization of up to 128 and 256 droplet generators on planar glass chips to achieve throughput of 128.0 and 320.0 mL/h of Janus droplets and single emulsions, respectively [14]. Microgrooves are etched onto synthesis silica glass substrate by DRIE to form different droplet generator geometries such as cross-junction, Y-shape co-flow, and even three-consecutive cross-junctions, joint in a circular manner for the production of various emulsions [13, 14]. The etched chip is sealed with another glass substrate, and the microfluidic chip is mounted on a stainless-steel holder connected to the inlets of different liquids. Some of the designs are illustrated in **Figure 8**, while **Figure 9** shows the arrangement of commercialized modules sharing a single set of infusion pumps, for generation of Janus microparticles of a few hundred kilograms per month [36].

Other parallelization devices have also been reported, especially for mass production of single-phase emulsions. The channels in the parallelization devices are commonly distributed in a ladderlike geometry, in a treelike branched geometry, or in a row, while different layers are needed for the introduction of all liquid phases through through holes to be distributed into the microchannels [35, 36]. Apart from DRIE, parallel channels can also be micromachined onto PMMA substrate, replicated onto PDMS via lithography, or 3D printed [37, 38]. A PDMS millipede device is also reported, with more than 500 nozzles arranged in 2 rows, and droplets are formed through the nozzles and break under static instability [39]. Jeong et al. demonstrated the mass production of W/O emulsions at kilo-scale using 1000 parallel flow-focusing drop generators arranged in a 20 × 50 dense array on a PDMS device which is currently the highest degree of parallelization ever achieved (see **Figure 10**) [40]. Currently only Nisisako's team has demonstrated mass-production of Janus

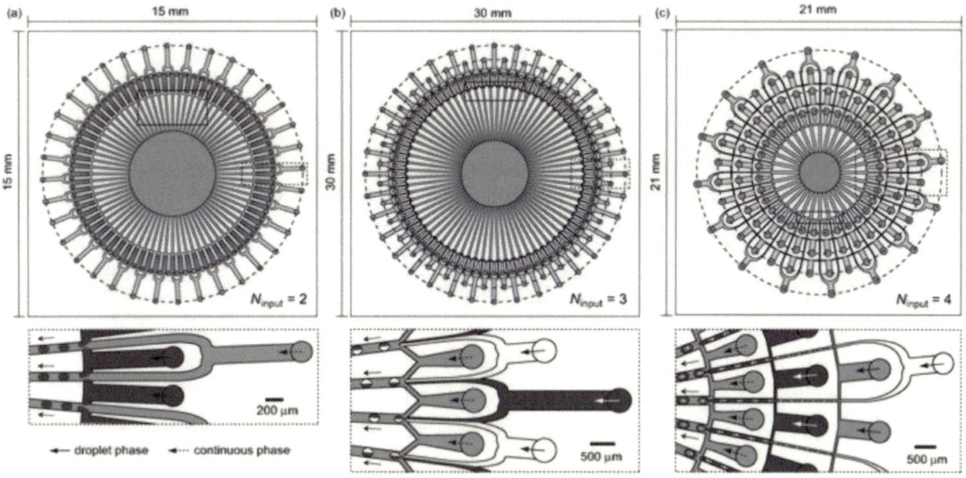

Figure 8.
Designs of glass microfluidic chips with parallelized channels in circular arrangements. (a) 144 cross-flowing droplet generators on two concentric circles, (b) 72 droplet generators on three concentric circles, and (c) 32 emulsion generators in four concentric circles producing triple emulsions. Reproduced from Ref. [13] with permission from the Royal Society of Chemistry.

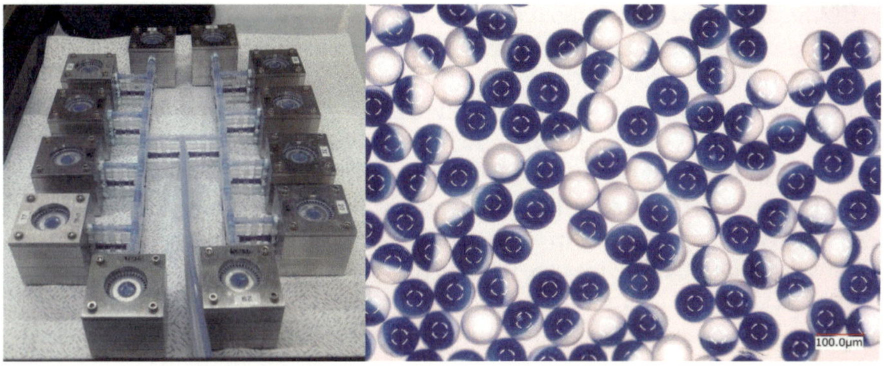

Figure 9.
Parallelized devices are arranged in series to further scale up the production of Janus particles. Reproduced with permission [36].

particles for commercial applications; however, there is no other information on the status of the parallelized device, whether the technology has been commercialized.

4.2 From "proof of concept" to commercialization

Inkjet printing is a classic example of commercialized microfluidic application to date, having been developed in the 1950s and largely commercialized and manufactured in the 1970s for digital image printing, through impelling droplets onto papers. The inkjet technology is subsequently adopted as a patterning tool for material deposition onto a wider range of substrates, for applications such as direct printing of electronic and optical devices [41]. More recently, studies have been conducted on the synthesis of microparticles via emulsion solvent evaporation following inkjet printing of droplets on substrates [42]. Following the debut of inkjet printing, notably only a limited number of microfluidic applications successfully made their way into the market in the following years, mainly in the areas of microfluidic genotyping and POC diagnostics. Chin et al. and Volpatti et al. summarized the commercialization of

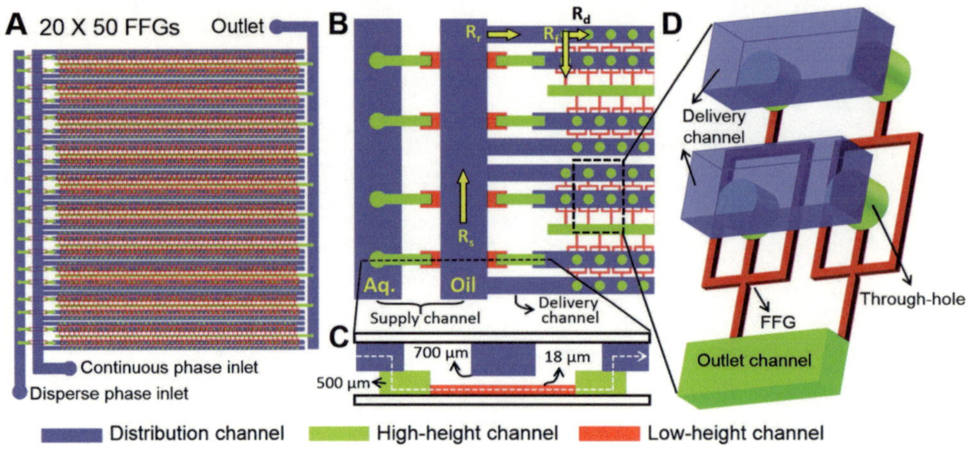

Figure 10.
Schematic diagram of a PDMS parallelization device. (A) the channels are arranged in a 20 × 50 array. (B) The device has a ladderlike geometry, and solution is distributed evenly across the device. (C) Cross-section view of the 3D microchannel. (D) Schematic of the distribution of solution from the delivery channel to the through holes and then the droplet generator; droplets are formed and then exit through the outlet channel. Reproduced from Ref. [40] with permission from the Royal Society of Chemistry.

microfluidic devices of which most are LOC-based POC diagnostic devices, reportedly commercialized by a few companies [5, 43]. These POC devices developed by individual companies cover a vast range of purposes, ranging from flu, cardiovascular diseases, malaria and *E. coli* detection to blood chemistry analysis. Devices for cancer detection and DNA and RNA signatures are still under development. These devices come in different forms such as disposable cards or cartridges containing preloaded test solutions and capillary-driven test strips, and the entire analysis could be performed on-chip or off-chip with handheld analyzers or benchtop instruments. More conventional POC devices are expected to adapt LOC concepts and functions in order to improve diagnosis and to meet greater clinical needs.

Alas, the overwhelming accomplishments in microfluidic research and development are not reflected in the slow uptake of microfluidic technology in the market. The industrial realization of microfluidic innovations is most often held back by the inflexibility of microfluidic components for integration, economic non-viability, and the lack of standardization. A complete microfluidic system usually entails many different fluidic components, and a single manufacturing process is not possible at current stage without the intervention and support from manual labor. Automated and assembly-free prototyping of simple devices is perceived to be a goal that is not attainable in the near future. Most importantly, with many deemed "promising" innovations, the challenge is to ensure that such innovations can be adapted and integrated to other existing appliances, technologies, or systems. Chin et al., Volpatti et al., and Mohammed et al. have all echoed the same concern that the focus should be on practical, marketable devices that are readily integrable, instead of unending development of single microfluidic component with niche function. Such is the demerit of introducing microfluidic-formulated solutions; they are too unique to be adapted; hence, the issue of chip-to-world interface should also be addressed and resolved even from the initial design phase [44]. And unless a minimum volume demand is met and a reasonable margin is secured, it is both uneconomic and unsustainable for mass production of microdevices, a risk that no stakeholders would bear [45].

The translation of the technology seems more challenging as it appears that academic researchers are more interested in publishing their inventions rather than developing them into commercially viable products. Many successful inventions remain a "proof of concept" and ceased to be developed. In fact, many of the microfluidic systems are limited to "one design one application" which are applicable only to niche research areas, rendering them less appealing for further development. The voluminous lateral development of microfluidic researches and lack of standardization on methods and materials instead become the stumbling block for a standard microfluidic solution to reach mainstream uses, hence slowing the commercial uptake of the technology [5, 44].

5. Conclusion

Microfluidic-based researches are growing in an unprecedented pace with more research disciplines adopting the technology, techniques, and tools that are now much established. At present, novel microfluidic platforms offering diverse functionalities and solutions are surfacing each day, and there seems to be no limit to what the microscale technology could offer, be it in clinical technologies or the manufacturing sector. While researches are flourishing in the academic perspective, the ambiguity surrounding the commercialization of microfluidic applications did not go unnoticed. The interest of much of the microfluidic community is in lateral development of microfluidic solutions with only a small amount ongoing researches targeting at resolving technical challenges for innovation-to-market translation: the technologies and economy for mass productions. Despite the continuous effort and enthusiasm for more microfluidic solutions to be realized industrially, the path to commercialization and industrialization for more microfluidic solutions depends potently on the joint effort and commitment from the researching teams and their industrial counterparts.

Acknowledgements

This work is supported by Zhejiang Provincial Natural Science Foundation of China under General Programme with project code LY19E060001 and Ningbo Science and Technology Bureau under Service Industry S&T Programme with project code 2019F1030.

Author details

Maxine Yew[1], Kaiseng Koh[2] and Yong Ren[1]*

1 Department of Mechanical, Materials and Manufacturing Engineering, University of Nottingham Ningbo China, Ningbo, China

2 School of Engineering and Physical Sciences, Heriot-Watt University Malaysia, Putrajaya, Malaysia

*Address all correspondence to: yong.ren@nottingham.edu.cn

References

[1] Gañán-Calvo AM. Generation of steady liquid microthreads and micron-sized monodisperse sprays in gas streams. Physical Review Letters. 1998;**80**(2):285-288. DOI: 10.1103/PhysRevLett.80.285

[2] Thorsen T, Roberts RW, Arnold FH, Quake SR. Dynamic pattern formation in a vesicle-generating microfluidic device. Physical Review Letters. 2001;**86**(18):4163-4166. DOI: 10.1103/PhysRevLett..86.4163

[3] Simon MG, Lee AP. Microfluidic droplet manipulations and their applications. In: Day P, Manz A, Zhang Y, editors. Microdroplet Technology: Principles and Emerging Applications in Biology and Chemistry. New York, NY: Springer New York; 2012. pp. 23-50. DOI: 10.1007/978-1-4614-3265-4_2

[4] Shang L, Cheng Y, Zhao Y. Emerging droplet microfluidics. Chemical Reviews. 2017;**117**(12):7964-8040. DOI: 10.1021/acs.chemrev.6b00848

[5] Volpatti LR, Yetisen AK. Commercialization of microfluidic devices. Trends in Biotechnology. 2014;**32**(7):347-350. DOI: 10.1016/j.tibtech.2014.04.010

[6] Ren K, Zhou J, Wu H. Materials for microfluidic chip fabrication. Accounts of Chemical Research. 2013;**46**(11):2396-2406. DOI: 10.1021/ar300314s

[7] Anna SL. Droplets and bubbles in microfluidic devices. Annual Review of Fluid Mechanics. 2016;**48**(1):285-309. DOI: 10.1146/annurev-fluid-122414-034425

[8] Umbanhowar PB, Prasad V, Weitz DA. Monodisperse emulsion generation via drop break off in a coflowing stream. Langmuir. 2000;**16**(2):347-351. DOI: 10.1021/la990101e

[9] Utada AS, Lorenceau E, Link DR, Kaplan PD, Stone HA, Weitz DA. Monodisperse double emulsions generated from a microcapillary device. Science. 2005;**308**(5721):537. DOI: 10.1126/science.1109164

[10] Zhu P, Wang L. Passive and active droplet generation with microfluidics: A review. Lab on a Chip. 2017;**17**(1):34-75. DOI: 10.1039/C6LC01018k

[11] Ji Q, Zhang JM, Liu Y, Li X, Lv P, Jin D, et al. A modular microfluidic device via multimaterial 3D printing for emulsion generation. Scientific Reports. 2018;**8**(1):4791. DOI: 10.1038/s41598-018-22756-1

[12] Takeuchi S, Garstecki P, Weibel DB, Whitesides GM. An axisymmetric flow-focusing microfluidic device. Advanced Materials. 2005;**17**(8):1067-1072. DOI: 10.1002/adma.200401738

[13] Nisisako T, Ando T, Hatsuzawa T. High-volume production of single and compound emulsions in a microfluidic parallelization arrangement coupled with coaxial annular world-to-chip interfaces. Lab on a Chip. 2012;**12**(18):3426-3435. DOI: 10.1039/C2LC40245A

[14] Nisisako T, Torii T. Microfluidic large-scale integration on a chip for mass production of monodisperse droplets and particles. Lab on a Chip. 2008;**8**(2):287-293. DOI: 10.1039/b713141k

[15] Mukhopadhyay R. When PDMS isn't the best. Analytical Chemistry. 2007;**79**(9):3248-3253. DOI: 10.1021/ac071903e

[16] Huh D, Hamilton GA, Ingber DE. From 3D cell culture to organs-on-chips.

Trends in Cell Biology. 2011;**21**(12):745-754. DOI: 10.1016/j.tcb.2011.09.005

[17] Sochol RD, Sweet E, Glick CC, Wu S-Y, Yang C, Restaino M, et al. 3D printed microfluidics and microelectronics. Microelectronic Engineering. 2018;**189**:52-68. DOI: 10.1016/j.mee.2017.12.01

[18] Li S, Gong X, Mc Nally CS, Zeng M, Gaule T, Anduix-Canto C, et al. Rapid preparation of highly reliable PDMS double emulsion microfluidic devices. RSC Advances. 2016;**6**(31):25927-25933. DOI: 10.1039/C6RA03225G

[19] Guckenberger DJ, de Groot TE, Wan AMD, Beebe DJ, Young EWK. Micromilling: A method for ultra-rapid prototyping of plastic microfluidic. Lab on a Chip. 2015;**15**(11):2364-2378

[20] Gale KB, Jafek RA, Lambert JC, Goenner LB, Moghimifam H, Nze CU, et al. A review of current methods in microfluidic device fabrication and future commercialization prospects. Inventions. 2018;**3**(3):60. DOI: 10.3390/inventions3030060

[21] Iliescu C, Taylor H, Avram M, Miao J, Franssila S. A practical guide for the fabrication of microfluidic devices using glass and silicon. Biomicrofluidics. 2012;**6**(1):016505-016516. DOI: 10.1063/1.3689939

[22] Fiorini GS, Chiu DT. Disposable microfluidic devices: Fabrication, function, and application. BioTechniques. 2005;**38**(3):429-446. DOI: 10.2144/05383RV02

[23] Steinbacher JL, Lui Y, Mason BP, Olbricht WL, McQuade DT. Simplified mesofluidic systems for the formation of micron to millimeter droplets and the synthesis of materials. Journal of Flow Chemistry. 2012;**2**(2):56-62. DOI: 10.1556/JFC-D-11-00023

[24] Yew M, Ren Y, Koh KS, Sun C, Snape C, Yan Y. Synthesis of microcapsules for carbon capture via needle-based droplet microfluidics. Energy Procedia. 2019;**160**:443-450. DOI: 10.1016/j.egypro.2019.02.179

[25] Bottaro E, Nastruzzi C. "Off-the-shelf" microfluidic devices for the production of liposomes for drug delivery. Materials Science and Engineering: C. 2016;**64**:29-33. DOI: 10.1016/j.msec.2016.03.056

[26] Benson BR, Stone HA, Prud'homme RK. An "off-the-shelf" capillary microfluidic device that enables tuning of the droplet breakup regime at constant flow rates. Lab on a Chip. 2013;**13**(23):4507-4511. DOI: 10.1039/c3lc50804h

[27] Terray A, Hart SJ. "Off-the-shelf" 3-D microfluidic nozzle. Lab on a Chip. 2010;**10**(13):1729-1731. DOI: 10.1039/B927244E

[28] Wu P, Wang Y, Luo Z, Li Y, Li M, He L. A 3D easily-assembled micro-cross for droplet generation. Lab on a Chip. 2014;**14**(4):795-798. DOI: 10.1039/C3LC51126J

[29] Li T, Zhao L, Liu W, Xu J, Wang J. Simple and reusable off-the-shelf microfluidic devices for the versatile generation of droplets. Lab on a Chip. 2016. DOI: 10.1039/C6LC00967K

[30] Gal-Or E, Gershoni Y, Scotti G, Nilsson SME, Saarinen J, Jokinen V, et al. Chemical analysis using 3D printed glass microfluidics. Analytical Methods. 2019;**11**(13):1802-1810. DOI: 10.1039/C8AY01934G

[31] Vijayan S, Hashimoto M. 3D printed fittings and fluidic modules for customizable droplet generators. RSC Advances. 2019;**9**(5):2822-2828. DOI: 10.1039/C8RA08686A

[32] Macdonald NP, Cabot JM, Smejkal P, Guijt RM, Paull B, Breadmore MC. Comparing microfluidic performance

of three-dimensional (3D) printing platforms. Analytical Chemistry. 2017;**89**(7):3858-3866. DOI: 10.1021/acs.analchem.7b00136

[33] Song R, Abbasi MS, Lee J. Fabrication of 3D printed modular microfluidic system for generating and manipulating complex emulsion droplets. Microfluidics and Nanofluidics. 2019;**23**(7):92. DOI: 10.1007/s10404-019-2258-234

[34] Chou W-L, Lee P-Y, Yang C-L, Huang W-Y, Lin Y-S. Recent advances in applications of droplet microfluidics. Micromachines. 2015;**6**(9):1249-1271. DOI: 10.3390/mi6091249

[35] Jeong H-H, Issadore D, Lee D. Recent developments in scale-up of microfluidic emulsion generation via parallelization. Korean Journal of Chemical Engineering. 2016;**33**(6):1757-1766. DOI: 10.1007/s11814-016-0041-6

[36] Nisisako T. Recent advances in microfluidic production of Janus droplets and particles. Current Opinion in Colloid & Interface Science. 2016;**25**:1-12. DOI: 10.1016/j.cocis.2016.05.03

[37] Conchouso D, Castro D, Khan SA, Foulds IG. Three-dimensional parallelization of microfluidic droplet generators for a litre per hour volume production of single emulsions. Lab on a Chip. 2014;**14**(16):3011-3020. DOI: 10.1039/C4LC00379A

[38] Femmer T, Jans A, Eswein R, Anwar N, Moeller M, Wessling M, et al. High-throughput generation of emulsions and microgels in parallelized microfluidic drop-makers prepared by rapid prototyping. ACS Applied Materials & Interfaces. 2015;7(23):12635-12638. DOI: 10.1021-acsami.5b03969

[39] Amstad E, Chemama M, Eggersdorfer M, Arriaga LR, Brenner MP, Weitz DA. Robust scalable high throughput production of monodisperse drops. Lab on a Chip. 2016;**16**(21):4163-4172. DOI: 10.1039/C6LC1075J

[40] Jeong H-H, Yelleswarapu VR, Yadavali S, Issadore D, Lee D. Kilo-scale droplet generation in three-dimensional monolithic elastomer device (3D MED). Lab on a Chip. 2015;**15**(23):4387-4392. DOI: 10.1039/C5LC01025J41

[41] Tekin E, Smith PJ, Schubert US. Inkjet printing as a deposition and patterning tool for polymers and inorganic particles. Soft Matter. 2008;**4**(4):703-713. DOI: 10.1039/B711984D

[42] Wang Y, Deng R, Yang L, Bain CD. Fabrication of monolayers of uniform polymeric particles by inkjet printing of monodisperse emulsions produced by microfluidics. Lab on a Chip. 2019;**19**(18):3077-3085. DOI: 10.1039/C9LC00588A

[43] Chin CD, Linder V, Sia SK. Commercialization of microfluidic point-of-care diagnostic devices. Lab on a Chip. 2012;**12**(12):2118-2134. DOI: 10.1039/C2LC21204H

[44] Mohammed MI, Haswell S, Gibson I. Lab-on-a-chip or Chip-in-a-lab: Challenges of commercialization lost in translation. Procedia Technology. 2015;**20**:54-59. DOI: 10.1016/j.protcy.2015.07.010

[45] Christian H. Large-scale droplet production in microfluidic devices—An industrial perspective. Journal of Physics D: Applied Physics. 2013;**46**(11):114008. DOI: 10.1088/0022-3727/11/114008

Chapter 3

Microfluidic Synthesis of Functional Materials as Potential Sorbents for Water Remediation and Resource Recovery

Voon-Loong Wong, Chin-Ang Isaac Ng, Lui-Ruen Irene Teo and Ci-Wei Lee

Abstract

The advance of droplet-based microfluidics has enabled compartmentalization and controlled manipulation of monodispersed emulsions with high yield and incorporation efficiency. It has become a highly exotic platform in synthesizing functional material due to the presence of two immiscible liquids and the interface between them. With its intrinsic feature in high degree of product control, advanced emulsion-based synthesis of functional material is constituted as a template for effective water remediation and resource recovery. This chapter aims to provide an overview of recent advances in microfluidic technology for environmental remediation. More specifically, the facility of microemulsion-based functional materials for water remediation is reviewed. Moreover, the removal and recovery of pollutants, such as heavy metal, dye, pharmaceuticals, etc., from aquatic environment by the applications of adsorption on functional micro/nanomaterials are unfolded with respect to its potential for wastewater purification.

Keywords: microfluidics, remediation, functional sorbents, recovery, wastewater

1. Introduction

Over the past few years, tremendous growth in the manufacturing and widespread application of synthetic chemical compounds in industrial sectors has led to growing number of emerging contaminants in environmental matrices (air, wastewater, water, sediment, and soil), which poses a challenge for regulatory agencies. Apart from that, the environment has gradually suffered as a result of exposure to the emerging pollutants especially in water and wastewater. Various types of physical, chemical, and biochemical methodologies have been reported on effectively removing pollutants. These methods include membrane separation, biological degradation, advanced oxidation process, and adsorption as well [1]. Recent developments have indicated that the adsorption method is favorable due to its wide availability, lower cost, as well as its recyclability. As in terms of cost effectiveness, simplicity of construction, and easy adaptation of operating conditions, particle adsorption materials become the key for realizing various adsorption applications in environment remediation using fixed bed reactors, absorption columns, fluidized beds, and cyclone separators [2]. However, controlled shapes, size, and compartments are some of limitations for most conventional methods. Unlike conventional approaches, microfluidics is comparably conducive as it has improved and extended the possibilities to synthesize highly controlled size of microparticles with excellent adsorption capability and reusability [2].

Microfluidics is a multidisciplinary field stretch across engineering, physics, chemistry, microtechnology, and biotechnology. Microfluidic devices generally have two of the three geometric length scales in the order of microns. The micrometer length scale defines the most obvious but extremely important character of microfluidic devices especially their small size, which allows small sample volumes, low cost, and fast analysis, but with high resolution and sensitivity [3]. With the length scale associated with microfluidic devices, the flow within them tends to be laminar. Moreover, one characteristic of microscale miniaturization is the large surface area to volume ratio. Thus, this favorable aspect plays a major role in control and manipulation of fluid flow in microfluidics. Recent advances and innovations could make microfluidics technology ubiquitous and create microfluidic devices that are more functional, efficient, and cost effective than conventional techniques.

The development of microfluidic systems that allow for the formation of microdroplets inside microfluidic devices has gained greatly attention over the past 20 years. The rise in interest is due to the utilization of microfluidic devices in a broad range of biological and biomedical application areas including disease diagnosis, cell treatment, drug screening, single-cell analysis, and drug delivery. Liquid droplets dispersed in a second immiscible fluid are useful, particularly when the sizes and the size distribution of droplet can be prescribed on a few hundred nanometers to a few millimeters [4]. Microfluidic emulsification approach offers an alternate and versatile route to produce emulsions that are highly monodispersed and have high formation frequencies in multiphase fluid systems [5]. Additionally, the geometrical attributes and flow characteristics within these microfluidics system constitutes flexibility in producing complex structured emulsions, such as double-emulsions and multi-emulsions [6, 7]. Highly monodispersed single, double, or multi-emulsions can be used as a template to prepare micro- and nanoparticles with various structures and morphologies [7], as shown in **Figure 1**. As seen in **Figure 1**, the presence of different particle shapes, compartments, and microstructures is formed based on the flow and geometrical attributes in microfluidics. Consequently, the superior properties of droplet-based microfluidic device have become extremely promising and attractive platform that enables the production of functionalized monodisperse microparticles.

In order to fabricate micro- or nanoparticles, the analysis of droplet formation is imperative to understand the device operation and its process control to meet different application purposes. Additionally, other's droplet manipulation, such as fusion, fission, mixing, and sorting with high precision and flexibility constitutes essential issue, at which extensive investigations have been directed. Based on the sources of the driving force involved, there are five approaches to droplet manipulation: hydrodynamic stress, electro-hydrodynamics, thermos-capillary, magnetism, and acoustics [8]. Hydrodynamic stress is a simple and effective approach to accomplish droplet manipulation relating to the geometrical characteristics of microchannel [9]. In this mode of manipulation, various methods have been employed in the formation of droplets in microfluidics device, including co-flowing mechanisms, flow-focusing mechanisms as well as cross-flowing mechanisms [10]. These mechanisms enabled the formation of dispersions with highly attractive features, particularly the control over droplet and particle size distribution.

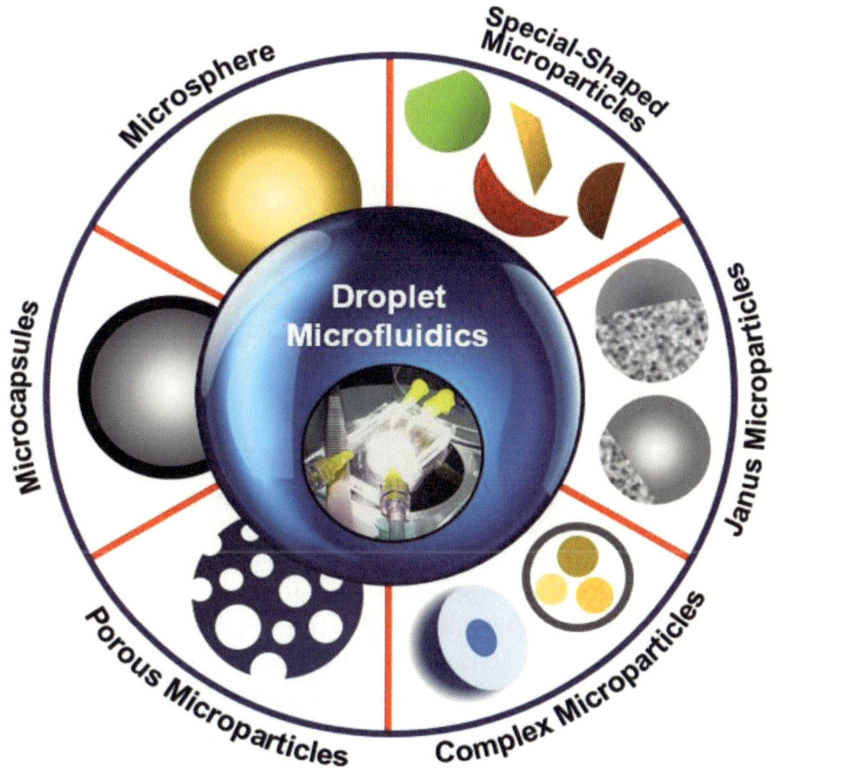

Figure 1.
Classification of various structures of microparticles produced by droplet-based microfluidics.

Over the years, microfluidic devices have been developed to synthesize particles for water remediation. For instance, Zhao et al. [11] synthesize graphene oxide microspheres using microfluidics technology for the removal of perfluorooctane sulfonate. Dong et al. studied the anionic dye adsorption using chitosan microparticles [2]. In addition, the performance of microcapsules for the CO_2 adsorption and permeability was investigated by Stolaroff et al. [12]. Hitherto, there are still little attempts that have been considered to use microfluidic platform as a selection scheme for large-scale industrial wastewater treatment. Toward practical and high capacity, microfluidic platforms generally suffer from the high cost and limited capacity for high throughput production of microfluidic synthesized particles. Nevertheless, it is possible to operate the production in parallel in order to realize continuous processes [13]. Moreover, the fundamental microfluidic research is still highly demanded to bridge the gap between the functional material synthesis and industrial perspective on exploring the possibilities and potential benefits of microfluidic processes [2, 13].

In this chapter, we will discuss the current trend of employing microfluidic technologies for environmental remediation, specifically for wastewater treatment and water remediation. Apart from that, we will also provide a general overview of the facility of microfluidics emulsification for the fabrication of various microparticles and nanoparticles as functional adsorbents. The sorption capacity and performance of the functional materials will be also evaluated in this review. Eventually, this chapter provides an impression of what are the consolidated fields of microfluidic formulation in functional material synthesis that will look like in about a decade from now.

2. Microfluidic technology for water remediation and resource recovery

Due to excessive discharges of harmful wastes and by-products to the environment, water contamination is the most prominent in which numerous organic and inorganic pollutants are found in the fresh water resources such as ponds, rivers, and underground water. Traditionally, there are various methods of water analysis including atomic adsorption, chemical analysis, chromatography, colorimetry, and spectrometry. Although these techniques have high sensitivity and accuracy, limitations such as expensive instrumentation, time consuming and requires manual operation in sampling process, causing the researchers to shift their interest toward microfluidics technology, which has a great potential to replace the traditional water analysis techniques [14].

2.1 Droplet microfluidic system

Droplet-based microfluidic technology involves the formation and manipulation of discrete droplets inside microdevices [15]. Droplet microfluidics is being widely used in different applications, such as chemical reactions, therapeutic agent delivery, imaging, biomolecule synthesis, diagnostic chips, and drug delivery [16]. The major concerns when creating droplet microfluidic device are the type of microfluidic chip fabrication materials used and the fluids used for droplet generation. Poly(dimethyl)siloxane (PDMS) is commonly used as fabrication material of microfluidic devices because PDMS is a relatively low cost and easily moldable elastomer. Nevertheless, PDMS has low solvent resistance, causing it to deform in the presence of strong organic solvents. Therefore, materials with higher solvent resistance, such as glass [17] and silicon, [18] are used. The pros and cons of the application of different microfluidic chip materials in droplet microfluidic system are shown in **Table 1**. The application of droplet microfluidics in the formation of functional materials for removal of pollutants will be discussed in Section 3 subsequently.

2.2 Microfluidic reactors

Microfluidic reactors, also known as microreactors, have been widely used in wastewater treatment because the development of microfluidics technology in this area helps to overcome some existing problems in bulk reactors. The two essential issues of bulk reactor are photon transfer limitations and mass transfer limitations [20]. Thus, great attention and interest have been shown in microfluidic system as microreactor inherits the merit of microfluidics. Recently, the technologies of microfluidics in advanced oxidation processes (AOPs) for wastewater treatment were studied and focused. AOPs involve the utilization

Application	Glass/ silicon	Elastomers	Thermoset	Thermoplastics	Hydrogel	Paper
Droplets formation	Excellent	Moderate	Good	Good	N/A	N/A
Production cost	High	Medium	High	Low	Medium to high	Low
Reusability	Yes	No	Yes	Yes	No	No

Table 1.
Pros and cons of different microfluidic chip materials for different applications [19].

of hydroxyl radicals ($OH^{\cdot}$) or sulfate radicals ($SO_4^{\cdot -}$) as a major oxidizing agent to effect water purification [21]. This is because these powerful radicals are extremely effective to destruct the organic and inorganic contaminants in wastewater and transform them to less or even non-toxic products [22]. The most popular AOPs that employ the technologies of microfluidics are photocatalysis and Fenton processes. The development and application of several AOP mechanisms will be discussed in the next sub-section. Then, a brief example of the application of different types of microreactors in water treatment and resource recovery is listed in **Table 2**.

2.2.1 Photocatalysis

Heterogenous photoassisted catalysis, known as photocatalysis is one of the examples of hydroxyl radical-based AOPs which is of particular concern in wastewater treatment. It involves the utilization of light for decomposition or mineralization of organic pollutants into innocuous product, such as carbon dioxide and water, in the presence of catalysts [31]. The application of microfluidic technology offers a great number of advantages in photocatalytic water treatment. Microfluidic structures have larger surface-area-to-volume ratio, typically in the range of 10,000–50,000 m^2/m^3 [32], compared to bulk reactor in which the surface-area-to-volume ratio is typically below 600 m^2/m^3 [33]. The surface-area-to-volume ratio can be much larger if nanoporous photocatalyst

Microfluidic device design	Application	Result	Ref.
Optofluidic planar reactor	Degradation of methylene blue	30% of dye degraded within 5 min with a reaction rate constant two orders higher than bulk reactor	[23]
Microcapillary reactor	Reduction of methylene blue	Reduction rate of dye increased by >150 times compared to batch system	[24]
Tree-branched centimeter-scale reactor	Degradation of volatile organic compounds	95% of pollutants (benzene, toluene, ethylbenzene, m-p xylenes and o-xylene) degraded in <5 s of residence time	[25]
Jet-aerated microfluidic flow-through reactor	Degradation of clopyralid as model organic pollutant	Clopyralid is eliminated effectively after 1 hour under several conditions	[26]
Microfluidic atmospheric-pressure plasma reactor	Degradation of methylene blue	>97% of dye degraded	[27]
Droplet microfluidic reactor	Extraction of lead (II)	Pb (II) ion was selectively and completely removed from the simulated wastewater effluent within 2.00 s	[28]
Microfluidic chip with polymethyl methacrylate (PMMA) plates	Adsorption of copper (II)	Adsorption capacity of 42.08 mg/g is achieved	[29]
Simple cross microchannel microfluidic device	Quantification of bacterial cells in potable water	Bacteria were accurately enumerated within 15 min after fluorescent staining	[30]

Table 2.
Application of microfluidic technology in water remediation and resource recovery.

film is used. Thus, higher heat transfer performance can be achieved by using microreactors. Furthermore, the rate of reaction is significantly increased and consequently favors having higher throughputs [34]. With the enhancement of reaction rate, the reaction time is reduced. The time taken for degradation process in a microreactor takes only several to tens of seconds [35], whereas bulk reactor requires several hours [36]. Besides, microfluidic layer has short diffusion length, typically 10–100 μm to ease the diffusion of organic pollutants to the reaction surface [31]. In addition, microreactors usually contain an immobilized photocatalyst film under the thin layer of fluid. This can ensure a uniform irradiation on the reaction surface, resulting in higher photon efficiency [37]. Microreactor has self-refreshing effect as the running fluid can refresh the reaction surface naturally. This helps to move away the reaction products and stabilize the photocatalysts. According to journal by Wang et al. [35], the photocatalysts in microreactor can hold several hundred runs of photocatalytic reactions, whereas the activity of photocatalysts in bulk reactors starts to degrade after 10 runs of reactions [38].

2.2.2 Other mechanisms in AOPs

Electro-Fenton process is an efficient AOP that involves activation of hydrogen peroxide by metal salts, typically iron, to produce hydroxyl radicals [39]. Electro-Fenton process is extremely effective in water remediation of the effluents, which cannot be efficiently treated using biological technologies [39–41]. Besides, AOPs with plasma-based water treatment (PWT) have been widely studied as PWT have the potential to reduce organic contaminants in wastewater. The application of microfluidic technology in PWT gives the benefits of large surface-area-to-volume ratio and flow control, in low-cost and portable devices [27]. However, more researches and development are needed to validate PWT performance at macro-scale [42].

3. Droplet microfluidics for the production of micro- or nanofunctional sorbents

Droplet-based microfluidics is formed through fabricating emulsions of uniform size. There are two approaches to produce emulsions, which are active and passive. Unlike actively controlled microfluidic devices, the breakup of discrete phase in continuous phase driven is controlled in a fully passive manner, which is caused by flow instabilities and hydrodynamic pressure without external actuations, such as mechanical, electrical, thermal, and magnetic method [43]. In this section, passive formation of emulsion-based microparticles is mainly discussed. Emulsions can be produced and manipulated with micro device of different geometries forming different sizes and morphologies [44]. Monodisperse emulsions are produced in laminar flow region and generated drop-by-drop where the over size, shape, and morphologies of micro droplets can be control precisely [45].

The superior properties of droplet microfluidics are advantageous for precise microparticle manufacturing for wastewater treatment. Micro and nanofunctional particles with various morphologies are prepared using templates. In general, microfluidic device forms highly monodisperse emulsified droplets and forms microparticles via solidification, while nanoparticles are formed using photochemical, chemicals or physical methods [46]. The space structure of the emulsions is controlled while preparing the adsorbent with microfluidic devices. By changing the device structure of microfluidic device, complex structure of droplets can be produced such as single, double, and multiple emulsion [17, 47]. These emulsion droplets are classified according to the structures and the evolvement of the emulsion into different structures and morphologies. Single emulsion can be a template to solid particles including spheres and non-spheres, while double and multiple emulsions can be templates for Janus particles, microcapsules, vesicles, hollow spheres or core-shell spheres [48].

3.1 Emulsion template: single, double or multiple

Single emulsions are droplets of one phase fluid dispersed in another immiscible phase fluid. Flow focusing, cross-flow, co-flow are systems that are frequently

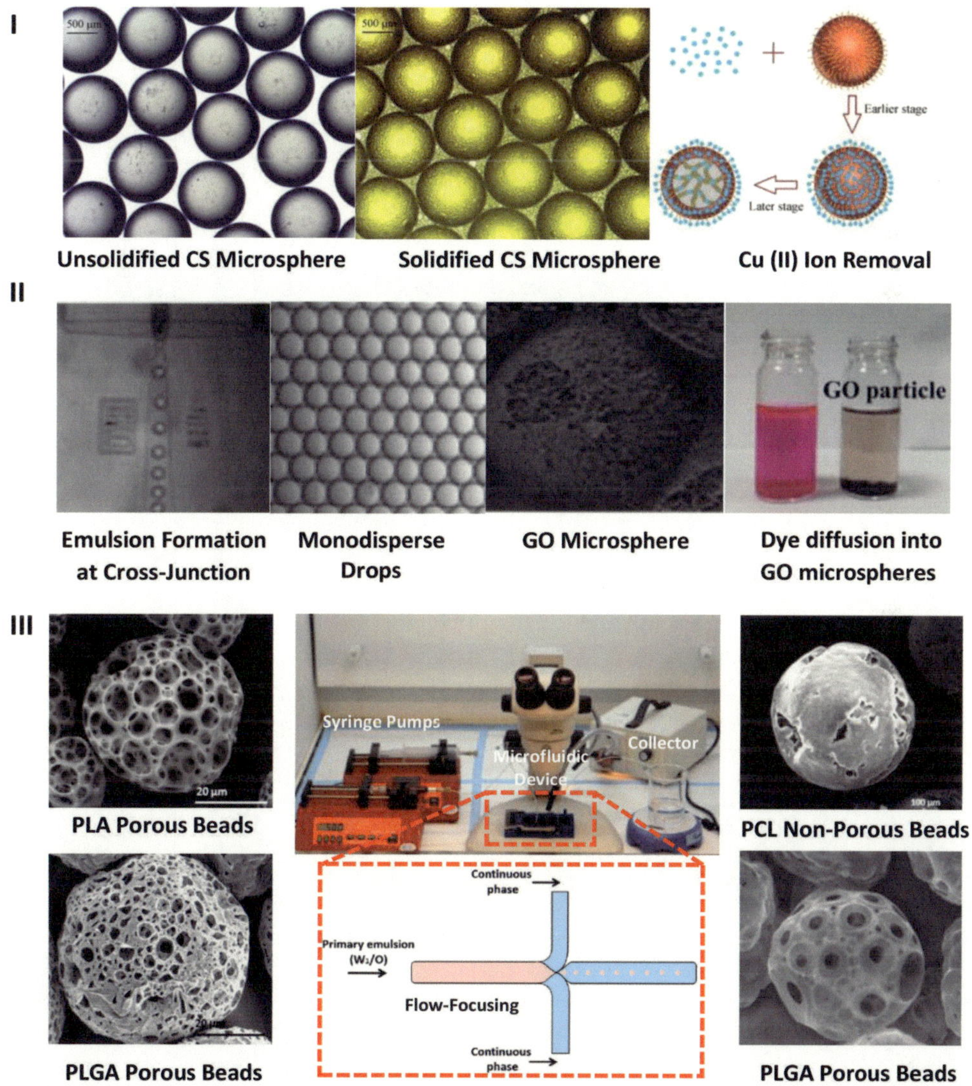

Figure 2.
Wastewater treatment by adsorption method onto microfluidic synthesized microparticles. (I) The adsorption process of heavy metal copper (II) ions using solidified chitosan microspheres (CS-MS) synthesized via cross-flowing PMMA-based microfluidics [29]. (II) Absorption of fluorescent dye molecules using PEGDA-graphene oxide (GO) microspheres [11]. (III) Microfluidic-based synthesis of porous and non-porous PLA, PLGA, and PCL microspheres [54]. (I) is reproduced with permission from Zhu et al., 2017, Copyright 2017, Springer Nature. (II) is reproduced with permission from Zhao et al. [11], Copyright 2016, Springer Nature. (III) is reproduced with permission from Amoyav and Benny [54], Copyright 2019, MDPI.

Microsorbent	Device	Approach	Channel size	Emulsion solidification	Ref.
Chitosan/polyethylenimine-chitosan microspheres	Commercial membrane	Membrane emulsification	Dispersed: 20 μm pores	Chemical cross-linking, solvent extraction	[55]
Chitosan/chitosan-poly(acrylic acid) composite microspheres	PMMA plates (laser fabrication)	Flow-focusing	Dispersed: 160 μm wide Continuous: 500 μm width	Chemical cross-linking, solvent extraction	[56]
Chitosan microspheres	PMMA plates (laser fabrication)	Cross-flowing	Dispersed: 1000 μm wide Continuous: 1000 μm width	Chemical cross-linking, solvent extraction	[29]
Chitosan microsphere	PMMA plates (laser fabrication)	Cross-flowing	Dispersed: 1000 μm width Continuous: 1000 μm width	Chemical cross-linking	[57]
Chitosan microparticles	PMMA plates (laser fabrication)	Flow-focusing	Dispersed: 500 μm width Continuous: 500 μm width	Chemical cross-linking	[2]
Thiourea-modified chitosan	PMMA plates (laser fabrication)	Flow-focusing	Dispersed: 160 μm width Continuous: 500 μm width	Chemical cross-linking, solvent extraction	[58]
Ion-imprinted chitosan microspheres	PMMA plates (laser fabrication)	Cross-flowing	Dispersed: 1000 μm width Continuous: 1000 μm width	Chemical cross-linking	[59]
Chitosan/silica hybrid microspheres	PMMA plates (laser fabrication)	Cross-flowing	Dispersed: 1000 μm width Continuous: 1000 μm width	Chemical cross-linking	[60]
Graphene oxide/$MgCl_2$-graphene oxide microspheres	PDMS chip (soft-lithography)	Cross-flowing	Not available	Photo-polymerization	[11]
Carbon nanotube microspheres	PDMS chip (3D printing/ soft-lithography)	Flow-focusing	Dispersed/Continuous: 900 μm width with a downstream constriction: 200 μm width	Decanting, solvent, cleaning, drying, pyrolysis	[61]
Hollow silica microspheres with ethyl butyrate	Glass microfluidic chip	Flow-focusing	Dispersed/continuous: 250 μm width	Hydrolyzation and condensation	[62]
PDMS microspheres	Off-the-shelf (needle-based) microfluidics	Flow-focusing	Dispersed: 510 μm width with downstream channel: 600 μm width	Photo-polymerization	[63]

Table 3.
Microfluidic synthesized spherical and functional microsorbents for water remediation and resource recovery.

used to form monodisperse droplets, and the coefficient of variation of droplets is usually less than 5% [3]. There are five breaking modes in passive generation, which are squeezing, dripping, jetting, tip-streaming, and tip-multi-breaking. These five modes have its own characteristics, for example, the structure and component of the droplets can be changed to produce inorganic nanoparticles, metal particles, and polymer particles [49]. Double or multi-emulsions are droplets with smaller droplets encapsulated in larger drops [48]. These emulsions are produced with capillary micro devices that involve three fluid streams in different capillaries. Initially, single droplets are formed when inner fluid is sheared by the middle fluid, then double or multiple droplets are formed when the outer fluid pinched off the single droplets containing in the middle fluid. There are difficulties of precisely controlling the shell thickness, aggregation, and secondary nucleation.

3.1.1 Special-shaped particles

Non-spherical particles have unique properties, and they are usually fabricated with many strategies such as seeded emulsion polymerization [50], template molding [51], and self-assembly [52]. However, a high quality, monodisperse, non-spherical particles with tailored geometries and shapes yet still difficult to produce using these methods. Droplets with different sizes and shapes in microfluidic channels are confined with microfluidic technologies for fabrication of non-spherical particles. The droplet will be deformed into ellipsoid, a disk, or a rod if the largest sphere can accommodate in the channel of a larger volume of droplet. Non-spherical droplet will be formed after they are solidified in the confined channel [53].

3.1.2 Spherical particles

There are different types of spherical particles, such as polymer microspheres, inorganic microspheres, noble metal nanospheres, and semiconductor nanospheres [48]. Polymer microspheres are usually prepared through spray-drying, coacervation, and emulsification [48]. Inorganic microsphere, which is the composed of titanium, silica, and carbon, have potential application in biomolecules, sensor, catalyst, and drug deliver. Noble metal nanosphere, such as gold, silver, and platinum, has shape and size dependent properties. It is hard to obtain desired size and size distribution as this individual nanoparticle tends to precipitate and coagulate to lower the surface energy. For the synthesis of semiconductor nanospheres, the microfluidic reactor should be chemically and thermally stable; thus, the droplets and carrier fluid could be stable, non-volatile, non-interacting, and immiscible from ambient to reaction temperatures. Hitherto, these microparticles are widely applicable in biological, pharmaceutical, medical (such as tumor treatment, drug controlled-release, and multi drug loading), optical, electrical applications, and researches. Moreover, polymer microspheres and inorganic microspheres are of great interest due to their potentials in adsorption separation as adsorbent in wastewater treatment, as shown in **Figure 2** and **Table 3** [29].

4. Sorption performance of different functional micro-sorbents for pollutants removal

Pollution management is now one of the most challenging issue facing modern societies. Due to the increasing population as well as industrialization of most countries, some pollutants are being discharged into aquatic environment without further treatment. This has a negative impact on the environment. The advancement of microfluidic technologies has allowed the synthesis of functional sorbents with greater sorption capacity. This is because the structure of the sorbent can be easily modified during emulsion. Thus, microsorbents with different functional groups can be synthesized to remove certain pollutants. Chitosan is a material that is widely used to make adsorbent for pollutant removal due to its affinity in removing heavy metals. In this subsection, the sorption performance of different functional microsorbents will be discussed as well as the kinetic model and adsorption isotherm.

4.1 Application and sorption performance of chitosan-based microsorbents

Chitosan is a natural polymer material that is found in abundance. It is made from the chitin of crustaceans and shrimps. Due to its affinity with heavy metals, it is a material with great potential for biosorbent synthesis. For instance, polyethylenimine-chitosan microspheres are used to remove methyl orange and Congo red dyes. Based on empirical observations, the uptake of methyl orange dye ranges around 88–97%, whereas the uptake of Congo red dye ranges around 86–96% [55]. Zhai et al. [57] and Dong et al. [2] both have reported that the sorption performance of chitosan microspheres in the uptake of the common textile azo dyes. Apart from synthetic dyes, chitosan microspheres were also synthesized to remove copper (II) ions, the sorption performance was observed to be roughly 38.52 mg/g [29]. With the addition of polyacrylic acid, the sorption performance increased significantly [56]. Besides, Lv et al. [64] studied the sorption uptake of copper (II) ion with using polyethylenimine-poly(glycidylmethacrylate)-chitosan microsphere. Microfluidic synthesized ion-imprinted chitosan microspheres and thiourea-modified chitosan were also used to remove copper (II) ions. These studies have proved that microfluidic synthesized chitosan is a promising biosorbent for water remediation. Moreover, its mechanical intensity, sorption performance, and equilibrium adsorption amount of emerging contaminants are highly enhanced as compared to those conventional methods.

4.2 Application and sorption performance of non-chitosan-based microsorbents

Aside from chitosan, other microfluidic synthesized materials such as graphene oxide, silicon-based organic polymer, carbon, and silica were also used to be studied for wastewater treatment [11, 61–63, 65]. Copic et al. [61] reported that sodium dodecyl sulfate (SDS) and Congo red dye can be removed using carbon nanotubes microspheres. Li et al. [62] investigated that the synthesized hollow silica microspheres offer a much higher storage capacity as compared to conventional hollow nanospheres. Moreover, the drug detoxification capability of the hollow silica microspheres containing ethyl butyrate was tested on iodine removal [62]. The sorption capacity of iodine by silica microspheres goes up to 95% removal uptake. Lian et al. [63] studied the removal of toluene using polydimethylsiloxane microspheres synthesized via needle-based microfluidic devices. Ren et al. [65] synthesized anisotropic Janus microparticles loaded with Fe_3O_4 and MnO_2 nanoparticles for the adsorption of basic dyes in wastewater. The sorption performance of the dye uptake using Janus micromotors can be varied from 47 to 94%. Thus, the current adsorption isotherm, kinetic modeling studies, and sorption performance of each microfluidics generated micro-sorbents for different water applications have been listed in **Table 4**.

Microsorbents	Application	Sorption performance	Adsorption isotherm	Kinetic modeling	Ref.
Chitosan/polyethylenimine-chitosan microspheres	Adsorption of methyl orange (MO) and Congo red (CR)	MO uptake: 88–97% CR uptake: 86–96%	Not available	Pseudo-second-order	[55]
Polyethylenimine-poly(glycidyl methacrylate)-chitosan microspheres	Adsorption of copper (II) ions	q_{max}: 229 mg/g	Langmuir	Pseudo-second-order	[64]
Chitosan/chitosan-poly(acrylic acid) composite microspheres	Adsorption of copper (II) ions	Chitosan: q_{max}: 50 mg/g chitosan-poly(acrylic acid): q_{max}: 66–72 mg/g	Langmuir	Pseudo-second-order	[56]
Chitosan microspheres	Adsorption of copper (II) ions	q_{max}: 38.52 mg/g	Langmuir	Pseudo-second-order	[29]
Chitosan microsphere	Adsorption of methyl orange	q_{max}: 207 mg/g	Langmuir	Pseudo-second-order	[57]
Chitosan microparticles	Adsorption of methyl orange	MO uptake: 15% to >95% (dosage from 1 to 7 mg). q_{max}: 182 mg/g (40 ppm)	Langmuir	Pseudo-second-order	[2]
Thiourea-modified chitosan	Adsorption of heavy metal copper (II) ions	q_{max}: 60.6 mg/g (100 ppm)	Not available	Pseudo-second-order	[58]
Ion-imprinted chitosan microspheres	Adsorption of heavy metal copper (II) ions	q_{max}: 81.97 mg/g (Ce < 400 ppm)	Langmuir	Pseudo-second-order	[59]
Chitosan/silica hybrid microspheres	Adsorption of heavy metal copper (II) ions	q_{max}: 53 mg/g (100 ppm)	Not available	Not available	[60]
Graphene oxide/$MgCl_2$-graphene oxide microspheres	Adsorption of perfluorooctane sulfonate (PFOS)	PFOS uptake: >95–98% (in 2 min) q_{max}: 5300 mg/g	Not available	Not available	[11]
Carbon nanotube microspheres	Adsorption of sodium dodecyl sulfate (SDS) and congo red (CR)	Filtration efficiency: CR (84.7%), SDS (61.6%)	Not available	Not available	[61]

Microsorbents	Application	Sorption performance	Adsorption isotherm	Kinetic modeling	Ref.
Hollow silica microspheres with ethyl butyrate	Waste removal and drug detoxification/ iodine removal	Iodine uptake: 95% (in 30 seconds)	Not available	Not available	[62]
Poly(dimethylsiloxane) microspheres	Adsorption of toluene	Toluene uptake: 30–45% (350 ppm)	Not available	Not available	[63]
Poly-(ETPTA) /Fe_3O_4/MnO_2 bubble-propelled micromotors	Adsorption of methylene blue (MB)	MB uptake: 47–94% (6.67 ppm in 6.7% w/w H_2O_2)	Not available	Not available	[65]

Table 4.
Adsorption isotherm and kinetic modeling studies of microfluidics generated micro-sorbents on the uptake of each emerging contaminants.

5. Conclusion

Certainly, microfluidic technologies are a relatively new research with great potential for development to enable more cost-effective synthesis of functional sorbents. Currently, the state-of-the-art microfluidic reactors for water remediation and resource recovery are being implemented in small-scale applications. Example of microfluidic reactors includes microfluidic atmospheric pressure plasma reactor, which is used to degrade methylene blue dyes. Furthermore, the advances in microfluidic technologies have improved the production of micro sorbents using microfluidic technology. Many efforts are also pouring into researching different functional sorbents for removal of different pollutants in wastewater. The results of such research have culminated in the discovery the adsorption capacity and isotherm model of different micro sorbents, which gives insight on the suitable sorbents required for different pollutants.

At present, the need of highly effective water remediation and resource recovery has brought about mounting interest in the research of microfluidic technologies. Many breakthroughs had been achieved in such research which enabled highly specific and controlled synthesis of functional sorbents using microfluidic systems. However, there are still many barriers that prevent the implementation of microfluidic technologies on an industrial scale. The greatest challenge against microfluidic technologies is the economical aspect. Due to the highly specific characteristics of sorbents used in wastewater treatment, most microfluidic systems are highly specific as well. Sorbents used in different applications would require different microfluidic systems to synthesize. Thus, more research will need to be done in order for such problems to be overcome. Of course, environmental problem such as water remediation and resource recovery cannot be solved by microfluidic technology alone and will need other technologies to complement it.

Acknowledgements

We gratefully acknowledge the funding support from Heriot-Watt Malaysia Campus.

Conflict of interest

The authors declare that they have no conflicts of interest.

Author details

Voon-Loong Wong*, Chin-Ang Isaac Ng, Lui-Ruen Irene Teo and Ci-Wei Lee
School of Engineering and Physical Sciences, Heriot-Watt University Malaysia Campus, Putrajaya, Malaysia

*Address all correspondence to: v.wong@hw.ac.uk

References

[1] Zhao J, Zou Z, Ren R, Sui X, Mao Z, Xu H, et al. Chitosan adsorbent reinforced with citric acid modified β-cyclodextrin for highly efficient removal of dyes from reactive dyeing effluents. European Polymer Journal. 2018;**108**:212-218. DOI: 10.1016/j.eurpolymj.2018.08.044

[2] Dong Z, Xu H, Bai Z, Wang H, Zhang L, Luo X, et al. Microfluidic synthesis of high-performance monodispersed chitosan microparticles for methyl orange adsorption. RSC Advances. 2015;**5**(95):78352-78360. DOI: 10.1039/C5RA17226H

[3] Zhao CX, Middelberg AP. Two-phase microfluidic flows. Chemical Engineering Science. 2011;**66**(7):1394-1411. DOI: 10.1016/j.ces.2010.08.038

[4] Anna SL, Bontoux N, Stone HA. Formation of dispersions using "flow focusing" in microchannels. Applied Physics Letters. 2003;**82**(3):364-366. DOI: 10.1063/1.1537519

[5] Sang L, Hong Y, Wang F. Investigation of viscosity effect on droplet formation in T-shaped microchannels by numerical and analytical methods. Microfluidics and Nanofluidics. 2009;**6**(5):621-635. DOI: 10.1007/s10404-008-0329-x

[6] Vladisavljević G, Al Nuumani R, Nabavi S. Microfluidic production of multiple emulsions. Micromachines. 2017;**8**(3):75. DOI: 10.3390/mi8030075

[7] Chen CH, Shah RK, Abate AR, Weitz DA. Janus particles templated from double emulsion droplets generated using microfluidics. Langmuir. 2009;**25**(8):4320-4323. DOI: 10.1021/la900240y

[8] Zhu P, Wang L. Passive and active droplet generation with microfluidics: A review. Lab on a Chip. 2017;**17**(1):34-75. DOI: 10.1039/C6LC01018K

[9] Torino S, Iodice M, Rendina I, Coppola G. Microfluidic technology for cell hydrodynamic manipulation. AIMS Biophysics. 2017;**4**(2):178-191. DOI: 10.3934/biophy.2017.2.178

[10] Wong VL, Loizou K, Lau PL, Graham RS, Hewakandamby BN. Numerical studies of shear-thinning droplet formation in a microfluidic T-junction using two-phase level-SET method. Chemical Engineering Science. 2017;**174**:157-173. DOI: 10.1016/j.ces.2017.08.027

[11] Zhao C, Fan J, Chen D, Xu Y, Wang T. Microfluidics-generated graphene oxide microspheres and their application to removal of perfluorooctane sulfonate from polluted water. Nano Research. 2016;**9**(3):866-875. DOI: 10.1007/s12274-015-0968-7

[12] Stolaroff JK, Ye C, Oakdale JS, Baker SE, Smith WL, Nguyen DT, et al. Microencapsulation of advanced solvents for carbon capture. Faraday Discussions. 2016;**192**:271-281. DOI: 10.1039/C6FD00049E

[13] Holtze C. Large-scale droplet production in microfluidic devices—An industrial perspective. Journal of Physics D: Applied Physics. 2013;**46**(11):114008. DOI: 10.1088/0022-3727/46/11/114008

[14] Liu Y, Jiang X. Why microfluidics? Merits and trends in chemical synthesis. Lab on a Chip. 2017;**17**(23):3960-3978. DOI: 10.1039/C7LC00627F

[15] Belder D. Microfluidics with droplets. Angewandte Chemie International Edition. 2005;**44**(23):3521-3522. DOI: 10.1002/anie.200500620

[16] Teh S, Lin R, Hung L, Lee AP. Droplet microfluidics. Lab on a Chip. 2008;**8**(2):198-220. DOI: 10.1039/B715524G

[17] Utada S, Lorenceau E, Link DR, Kaplan PD, Stone HA, Weitz DA. Monodisperse double emulsions generated from a microcapillary device. Science. 2005;**308**(5721):537-541. DOI: 10.1126/science.1109164

[18] Pollack MG, Shenderov AD, Fair RB. Electrowetting-based actuation of droplets for integrated microfluidics. Lab on a Chip. 2002;**2**(2):96-101. DOI: 10.1039/B110474H

[19] Ren K, Zhou J, Wu H. Materials for microfluidic chip fabrication. Accounts of Chemical Research. 2013;**46**(11):2396-2406. DOI: 10.1021/ar300314s

[20] Carp O, Huisman CL, Reller A. Photoinduced reactivity of titanium dioxide. Progress in Solid State Chemistry. 2004;**32**(1-2):33-177. DOI: 10.1016/j.progsolidstchem.2004.08.001

[21] Deng Y, Zhao R. Advanced oxidation processes (AOPs) in wastewater treatment. Current Pollution Reports. 2015;**1**(3):167-176. DOI: 10.1007/s40726-015-0015-z

[22] Huang CP, Dong C, Tang Z. Advanced chemical oxidation: Its present role and potential future in hazardous waste treatment. Waste Management. 1993;**13**(5-7):361-377. DOI: 10.1016/0956-053X(93)90070-D

[23] Lei L, Wang N, Zhang XM, Tai Q, Tsai DP, Chan HLW. Optofluidic planar reactors for photocatalytic water treatment using solar energy. Biomicrofluidics. 2010;**4**(4):043004. DOI: 10.1063/1.3491471

[24] Li X, Wang H, Inoue K, Uehara M, Nakamura H, Miyazaki M, et al. Modified micro-space using self-organized nanoparticles for reduction of methylene blue. Chemical Communications. 2003;(8):964-965. DOI: 10.1039/B300765K

[25] Azzouz I, Habba YG, Capochichi GM, Marty F, Vial J, Leprince WY, et al. Zinc oxide nano-enabled microfluidic reactor for water purification and its applicability to volatile organic compounds. Microsystems & Nanoengineering. 2018;**4**:17093. DOI: 10.1038/micronano.2017.93

[26] Pérez JF, Llanos J, Sáez C, López C, Cañizares P, Rodrigo MA. On the design of a jet-aerated microfluidic flow-through reactor for wastewater treatment by electro-Fenton. Separation and Purification Technology. 2019;**208**(8):123-129. DOI: 10.1016/j.seppur.2018.04.021

[27] Patinglag L, Sawtell D, Iles A, Melling LM, Shaw KJ. A microfluidic atmospheric-pressure plasma reactor for water treatment. Plasma Chemistry and Plasma Processing. 2019;**39**(3):561-575. DOI: 10.1007/s11090-019-09970-z

[28] Kurniawan YS, Sathuluri RR, Iwasaki W, Morisada S, Kawakita H, Ohto K, et al. Microfluidic reactor for Pb (II) ion extraction and removal with an amide derivative of calix[4]arene supported by spectroscopic studies. Microchemical Journal. 2018;**142**:377-384. DOI: 10.1016/j.microc.2018.07.001

[29] Zhu Y, Bai Z, Wang B, Zhai L, Luo W. Microfluidic synthesis of renewable biosorbent with highly comprehensive adsorption performance for copper (II). Frontiers of Chemical Science and Engineering. 2017;**11**(2):238-251. DOI: 10.1007/s11705-017-1627-1

[30] Sakamoto C, Yamaguchi N, Yamada M, Nagase H, Seki M, Nasu M. Rapid quantification of bacterial cells in potable water using

a simplified microfluidic device. Journal of Microbiological Methods. 2006;**68**(3):643-647. DOI: 10.1016/j.mimet.2006.11.003

[31] Wang N, Zhang X, Wang Y, Yu W, Chan HLW. Microfludic reactors for photocatalytic water purification. Lab on a Chip. 2014;**14**(6):1074-1082. DOI: 10.1039/C3LC51233A

[32] Yildiz OE, Yesil CO. Diffusion phenomena of cells and biomolecules in microfluidic devices. Biomicrofluidics. 2015;**9**(5):052606. DOI: 10.1063/1.4923263

[33] Chen YL, Kuo L, Tseng ML, Chen HM, Chen C, Huang HJ, et al. ZnO nanorod optical disk photocatalytic reactor for photodegradation of methyl orange. Optics Express. 2013;**21**(6):7240-7249. DOI: 10.1364/OE.21.007240

[34] Nguyen N, Wereley ST, Shaegh SAM. Fundamentals and Applications of Microfluidics. 3rd ed. Norwood: Artech House; 2018

[35] Wang N, Zhang X, Chen B, Song W, Chan NY, Chan HLW. Microfluidic phtoelectrocatalytic reactors for water purification with an integrated visible-light source. Lab on a Chip. 2012;**12**(20):3983-3990. DOI: 10.1039/C2LC40428A

[36] Oelgemoeller M. Highlights of photochemical reactions in microflow reactors. Chemical Engineering & Technology. 2012;**35**(7):1144-1152. DOI: 10.1002/ceat.201200009

[37] Herrmann JM. Heterogeneous photocatalysis: Fundamentals and applications to the removal of various types of aqueous pollutants. Catalysis Today. 1999;**53**(1):115-129. DOI: 10.1016/S0920-5861(99)00107-8

[38] Wu JCS, Wu T, Chu T, Huang H, Tsai D. Application of optical-fiber photoreactor for CO2 photocatalytic reduction. Topics in Catalysis. 2008;**47**(3-4):131-136. DOI: 10.1007/s11244-007-9022-7

[39] Brillas E, Sirés I, Oturan MA. Electro-Fenton process and related electrochemical technologies based on Fenton's reaction chemistry. Chemical Reviews. 2009;**109**(12):6570-6631. DOI: 10.1021/cr900136g

[40] Moreira FC, Boaventura RAR, Brillas E, Vilar VJP. Electrochemical advanced oxidation processes: A review on their application to synthetic and real wastewaters. Applied Catalysis B: Environmental. 2017;**202**:217-261. DOI: 10.1016/j.apcatb.2016.08.037

[41] Ribeiro AR, Nunes OC, Pereira MFR, Silva AMT. An overview on the advanced oxidation processes applied for the treatment of water pollutants defined in the recently launched directive 2013/39/EU. Environment International. 2015;**75**:33-51. DOI: 10.1016/j.envint.2014.10.027

[42] Foster JE. Plasma-based water purification: Challenges and prospects for the future. Physics of Plasmas. 2017;**24**(5):055501. DOI: 10.1063/1.4977921

[43] Shang L, Cheng Y, Zhao Y. Emerging droplet microfluidics. Chemical Reviews. 2017;**117**(12):7964-8040. DOI: 10.1021/acs.chemrev.6b00848

[44] Seemann R, Brinkmann M, Pfohl T, Herminghaus S, et al. Droplet based microfluidics. Reports on Progress in Physics. 2011;**75**(1):016601. DOI: 10.1088/0034-4885/75/1/016601

[45] Amstad E, Chen X, Eggersdorfer M, Cohen N, Kodger TE, Ren CL, et al. Parallelization of microfluidic flow-focusing devices. Physical Review E. 2017;**95**(4):043105. DOI: 10.1103/PhysRevE.95.043105

[46] Zhang H, Tumarkin E, Sullan RMA, Walker GC, Kumacheva E. Exploring microfluidic routes to microgels of biological polymers. Macromolecular Rapid Communications. 2007;**28**(5):527-538. DOI: 10.1002/marc.200600776

[47] Shah RK, Shum HC, Rowat AC, Lee D, Agresti JJ, Utada AS, et al. Designer emulsions using microfluidics. Materials Today. 2008;**11**(4):18-27. DOI: 10.1016/S1369-7021(08)70053-1

[48] Wang J, Li Y, Wang X, Wang J, Tian H, Zhao P, et al. Droplet microfluidics for the production of microparticles and nanoparticles. Micromachines. 2017;**8**(1):22. DOI: 10.3390/mi8010022

[49] Xu S, Nie Z, Seo M, Lewis P, Kumacheva E, Stone HA, et al. Generation of monodisperse particles by using microfluidics: Control over size, shape, and composition. Angewandte Chemie. 2005;**117**:734-738. DOI: https://doi.org/10.1002/ange.200462226

[50] Kim JW, Larsen RJ, Weitz DA. Uniform nonspherical colloidal particles with Tunable shapes. Advanced Materials. 2007;**19**(15):2005-2009. DOI: 10.1002/adma.200602345

[51] Rolland JP, Maynor BW, Euliss LE, Exner AE, Denison GM, DeSimone JM. Direct fabrication and harvesting of monodisperse, shape-specific Nanobiomaterials. Journal of the American Chemical Society. 2005;**127**(28):10096-10100. DOI: 10.1021/ja051977c

[52] Sacanna S, Pine DJ. Shape-anisotropic colloids: Building blocks for complex assemblies. Current Opinion in Colloid & Interface Science. 2011;**16**(2):96-105. DOI: https://doi.org/10.1016/j.cocis.2011.01.003

[53] Dendukuri D, Tsoi K, Hatton TA, Doyle PS. Controlled synthesis of nonspherical microparticles using microfluidics. Langmuir. 2005;**21**(6):2113-2116. DOI: 10.1021/la047368k

[54] Amoyav B, Benny O. Microfluidic based fabrication and characterization of highly porous polymeric microspheres. Polymers. 2019;**11**(3):419. DOI: 10.3390/polym11030419

[55] Wolska J. Chitosan and chitosan-polyethyleneimine microspheres prepared by membrane emulsification and their application for drug delivery systems. Progress on Chemistry and Application of Chitin and its Derivatives. 2017;**22**:220-235. DOI: 10.15259/PCACD.22.22

[56] Xu J, Xu X, Zhao H, Luo G. Microfluidic preparation of chitosan microspheres with enhanced adsorption performance of copper (II). Sensors and Actuators B: Chemical. 2013;**183**:201-210. DOI: 10.1016/j.snb.2013.04.004

[57] Zhai L, Bai Z, Zhu Y, Wang B, Luo W. Fabrication of chitosan microspheres for efficient adsorption of methyl orange. Chinese Journal of Chemical Engineering. 2018;**26**(3):657-666. DOI: 10.1016/j.cjche.2017.08.015

[58] Zhao H, Xu J, Lan W, Wang T, Luo G. Microfluidic production of porous chitosan/silica hybrid microspheres and its Cu (II) adsorption performance. Chemical Engineering Journal. 2013;**229**:82-89. DOI: 10.1016/j.cej.2013.05.093q

[59] Zhu Y, Bai ZS, Wang HL. Microfluidic synthesis of thiourea modified chitosan microsphere of high specific surface area for heavy metal wastewater treatment. Chinese Chemical Letters. 2017;**28**(3):633-641. DOI: 10.1016/j.cclet.2016.10.031

[60] Zhu Y, Bai Z, Luo W, Wang B, Zhai L. A facile ion imprinted synthesis

of selective biosorbent for Cu2+ via microfluidic technology. Journal of Chemical Technology & Biotechnology. 2017;**92**(8):2009-2022. DOI: 10.1002/jctb.5193

[61] Copic D, Maggini L, De Volder M. Monodisperse CNT microspheres for high permeability and efficiency flow-through filtration applications. Advanced Materials. 2018;**30**(12):1706503. DOI: 10.1002/adma.201706503

[62] Li D, Guan Z, Zhang W, Zhou X, Zhang WY, Zhuang Z, et al. Synthesis of uniform-size hollow silica microspheres through interfacial polymerization in monodisperse water-in-oil droplets. ACS Applied Materials & Interfaces. 2010;**2**(10):2711-2714. DOI: 10.1021/am100593b

[63] Lian Z, Ren Y, He J, Chen GZ, Koh KS. Synthesis of polydimethylsiloxane microspheres using microfluidics for treatment of toluene in wastewater. In: ASME 2018 16th International Conference on Nanochannels, Microchannels, and Minichannels. American Society of Mechanical Engineers Digital Collection. 2018. pp. V001T12A006-V001T12A006

[64] Lv L, Zhang J, Yuan S, Huang L, Tang S, Liang B, et al. Enhanced adsorption of Cu (II) ions on chitosan microspheres functionalized with polyethylenimine-conjugated poly (glycidyl methacrylate) brushes. RSC Advances. 2016;**6**(81):78136-78150. DOI: 10.1039/C6RA16226F

[65] Ren M, Guo W, Guo H, Ren X. Microfluidic fabrication of bubble-propelled micromotors for wastewater treatment. ACS Applied Materials & Interfaces. 2019. DOI: 10.1021/acsami.9b05925

Chapter 4

Application of Microfluidics in Biosensors

Jing Wang, Yong Ren and Bei Zhang

Abstract

This chapter reviews the up-to-date researches in the field of biosensors integrated with microfluidic techniques, most of which are publications within the last 5 years. The features of these biosensors, their applications, challenges, and possible future research interests in this field are also reviewed.

Keywords: microfluidics, biosensor, bioreceptor, lab-on-a-chip, bioaffinity, PDMS, μPAD, paper-based microfluidics, electrochemistry, optics, surface plasmon resonance, colorimetric, fluorescent, cell culture, food safety

1. Introduction

Biosensors are defined, by Tudos and Schasfoort [1], as "analytical devices comprised of a biological element (tissue, microorganism, organelle, cell receptor, enzyme, antibody) and a physicochemical transducer. Specific interaction between the target analyte and the biological material produces a physico-chemical change detected by the transducer. The transducer then yields an analog electronic signal proportional to the amount (concentration) of a specific analyte or group of analytes" [1]. The features of biosensors include the bio-recognition unit and the transduction mechanism from biological signals to measurable electronic signals, e.g., color, current, voltage, capacitance, light intensity, wavelength, and phase. Major parameters to assess the performances of biosensors include the following:

- High sensitivity. The sensitivity of a biosensor is always the first and one of the most important parameters in assessing its performance. The efficiency in capturing analytes, the specific characterization of the analyte, the capability of converting biological signals into electronic signals (or the response of the system), and the systematic and environmental noises both determine the sensitivity of a biosensor.

- High stability and repeatability. The stability of a biosensor refers to the capability of a biosensor in performing consistently and reliably under designated environments, and the stability of a biosensor is especially important when assessing portable or wearable biosensors that usually are applied in scenarios involving varied temperature, velocity, humidity, pressure, lighting conditions, etc. The repeatability of a biosensor mainly refers to the long-term performance of a biosensor under the same conditions and is usually tested regularly in commercial biosensors in order to recalibration.

- Quick response or real-time analysis and diagnosis. Real-time analysis usually delivers more information than providing a final result, e.g., binding rate, reaction time, kinetics, and saturation conditions, which can serve for the analysis in the applications of biological and chemical reactions and drug analysis. Response time required to bind sufficient molecules upon the sensing surface is typically determined by diffusion, which can extend to hours and even to days to generate a signal above the background noise level. This applies fundamentally to all sensors that accumulate and concentrate target molecules onto a transducer, including fluorophore-tagged molecules in microarray spots, label-free optical biosensors, and impedance-based sensors [2].

- Low consumption of sample volume. The samples for biosensors are usually with low volume due to the nature, e.g., tissue, antibody, and some biological samples are with low concentration or small molecular weight, and this enforces biosensors to perform with minimum sample consumption.

- Ease of operation. The application of biosensors goes from laboratory-based research to commercially available devices for at-home use. The operation of biosensors should eliminate professional operation or understanding of the device, but simply involves collecting samples and reading results.

- High throughput. Single-function biosensors are fading away from stage even with extremely low cost. Biosensors should be able to integrate all the good qualities mentioned in above bullet points together with the capability of multiple-tasking.

In traditional clinical healthcare, interests are in high-quality biosensor for the measurement of physiological indices. With the development in the requests of Internet of Things and self-service techniques, the interests and emphasis in healthcare transfers from clinical healthcare to family healthcare, e.g., long-term monitoring of chronic disease [3], disease prevention and early detection, reachable clinical services for remote districts, etc. Under the precondition of high quality, biosensors miniaturized as portable or wearable are emerging to meet the new trend of era, which promise a bright future in health management and digital health; researches on the biological and instrumental parts of point-of-care (POC) and lab-on-a-chip (LOC) techniques belong to this area. The potential applications of biosensors include real-time health monitoring, remotely synchronizing health data with medical personnel, patient management, POC disease diagnosis, big data statistics in health management, etc. [4]. Real-time health monitoring in domestic applications enables patients to monitor the health status by themselves at home without the assistance of professional personnel at minimum cost. Health data synchronized with medical personnel spares the patients from transporting and waiting for outpatient services, saving time and reducing medical expenses. Based on the collected health data, a potential service of clinical diagnosis is possible in an artificial intelligent health system in the near future [5]. Meanwhile, these new applications impose more requirements on the researches and developments of biosensors, as stated below.

- Low cost. For at-home applications, lower cost of biosensors is vital, so there are emerging researches on adopting cheap materials, simplifying sensing systems and adopting smart phones in data processing, etc.

- Noninvasive collection of samples. Biosensors for sensing human samples, noninvasive sensing, is preferable, especially for everyday or frequent measurements. Researches using human samples as saliva, tear, sweat, and urine are one of the hottest topics.

- Miniaturization of the biosensing systems [4, 5], including sample pre-processing unit, sensing unit, data collection/processing system, and data displaying unit.

- The design of integrated sensor chip. A sensor chip with multiple functions is preferable especially for sensing human samples.

2. Challenges in biosensing technologies

All the qualities mentioned above, which both researchers and industry are seeking for, raise multiple challenges, and we try to summarize and list below:

- Specific binding. The recognition of analytes should be specific only to the analytes which is not affected by other chemicals, molecules, or cells. This is significantly more challenging when the sample components are complex and mixed with various kinds of molecules. For example, the detection of one specific antibody in human blood sample should eliminate the effects of all other antibodies, cells, electrolytes, etc., the detection reflects only the concentration of this antibody, and the detection of one specific heavy metal ion in a real polluted water sample should be able to distinguish the reaction induced by all other ions.

- Non-specific binding, i.e., biofouling, in some cases significantly introduces signal noise, drift, or delay in biosensors [6]. The most common method to reduce non-specific binding is to completely wash the binding surface with buffer after the binding processing is finished; thus the weak binding induced by non-specific binding could be eliminated to the minimum extent.

- Properties of bioreceptors, e.g., concentration and alignment. Plenty of papers [3, 7–10] presented the protocols of surface activation, modification, and functionalization, but the protocols greatly depend on operation and environment. Even following exactly the same protocol, the coverage rate of bioreceptors and the repeatability of operation may vary a lot, due to the immobilization of multiple layers on the sensor surface, which usually involves linking layers for stable sensor surfaces like gold and silicon, so the surface treatment protocols should be tested before operating the binding events. So far quite limited number of papers presented theoretical and/or experimental analysis on the effect of bioreceptor alignment/orientation on the performance of biosensors. The linking layer molecules are usually randomly polarized and oriented, which can induce destructive interference and dramatically reduce the collective charge polarization [11], and this means that even the surface treatment protocols could be repeated and the alignment of bioreceptors is another parameter that will highly affect the outcomes of the binding events. The detection becomes less sensitive. Chu [12] proposed a method to homogenize the orientation of the chemical linker on nanowire-based field-effect sensor by applying an external voltage on a metal plate about 1 mm above the chip surface at certain frequency while grounding the back gate electrode; thus the molecular conformation can be maintained for hours or longer, and this method has been tested and proven by the detection of DNA hybridization reactions with poly-15 T ssDNA, showing that the alignment process promotes the sensitivity by 10-fold.

- Design of biosensor assay matrix. The effects of specific and non-specific binding on signal was tested and analyzed by Schneider [13, 14], which proves that for all the binding events, proper design of the sensor assay should be optimized, especially when the sample components is complex, for example, proper reference binding sites should be included in order to eliminate non-specific binding from different components. But this in another way increased the requirement in both the imaging capability of the biosensor and the data processing capability. Meanwhile, the surface treatment, modification, and functionalization [4, 7, 9, 10, 15, 16] will be more complicated and need to be tested and verified, and the complexity in sensor surface properties (e.g., various refractive indices of bioreceptors) requests better system compatibility and responsivity.

- Low concentration target molecule within a low-volume sample, i.e., extremely limited number of analytes available for detection. For the example, in the research of POC and LOC, 20–50 μL finger prick blood contains over 20,000 kinds of biomarkers of clinical interest at concentrations as low as 10 pg./mL, meaning that only 106–107 available biomolecules for one target [2].

3. What is microfluidics?

The definition of microfluidics, given by Whitesides from Harvard University, is:

It is the science and technology of systems that process or manipulate small amounts (10^{-9} to 10^{-18} liters) of fluids, using channels with dimensions of tens to hundreds of micrometers. It offers fundamentally new capabilities in the control of concentrations of molecules in space and time [17].

The material most commonly adopted for the fabrication of microfluidic structures is poly-dimethylsiloxane (PDMS), which is optically transparent and able to support important microfluidic components, e.g., pneumatic valves; meanwhile there are other materials with research interests, e.g., polycarbonate, polyolefin, silicon, and glass. Recently paper-based (reviews by [3, 18–21], and research by [22–28]) and cloth-based [29] microfluidics are drawing more attention because of the low cost, easy fabrication, and lightweight which are essential properties for POC applications. The major features of microfluidics are the small consumption of liquid sample and tiny dimensions of structures, which have significant impact on the development of biosensors, so the integration of microfluidics into biosensing techniques complies with the development of the era and generates unique features in biosensors, e.g., trace level of sample at high sensitivity.

4. Advantages of microfluidic-integrated biosensors

Microfluidics provide a closed and stable biosensing environment so to improve sensitivity. For on-site portable biosensors, the effect of an open environment on sensing results hugely lowers the biosensor performance. By integrating the microfluidic structures, sample processing and biosensing reactions are carried out within a closed and relatively stable environment, thus promising better sensitivity and reliability [30]. Taking the example of the application of solid-phase polymerase chain reaction (SP-PCR) in online molecular diagnosis, the development of this technique is hindered by lack of sensitive and portable on-chip optical detection technology. Hung [24] proposed a LOC device which combined the solid-phase polymerase chain reaction with supercritical angle fluorescence (SAF) microlens array embedded in a microchip. He demonstrated a high sensitivity of 1.6 copies/μL and showed comparable detection limit and linear range to off-chip detection using conventional laser scanner, and he stated this device as an on-chip highly sensitive and multiplexed pathogen detection with low-cost and compact optical components.

Microfluidic channel can efficiently, accurately, and significantly reduce the sensing area. Simulations and experiment results have shown that reducing the sensing area could shorten sensing time and increase the sensitivity with smaller sample requirement, especially at lower target concentrations [7, 11]. An increase in sensitivity of two orders of magnitude has been reported by Li et al. [31]. Meanwhile, the distribution of binding events along the sensing surface could be heterogeneous, and this could be induced by the heterogeneous in bioreceptor coverage, different alignment of bioreceptors, nonuniform concentrations of targets within samples in laminar flow, and nonuniform temperature, pressure, or other physical parameters along the sensing surface, the heterogeneity can be eliminated to minimum. Reduced sensing area also means miniaturizing sample volume which is essentially valuable to low concentration targets and rare targets with limited access.

Microfluidic structures are capable of integrating multiple functions within one device without introducing extra equipment or tools. For example, by designing and optimizing microfluidic channels, sample injection, pretreatment, and processing can be easily realized. Usually for biosensing, the modification of the biosensing surface is compulsory for specific binding of targets, and this is doable in microfluidic structures which are even more stable and more promising than manual operations. For the biosensing events, the volume and speed control of sample are achievable which provides more valuable information, e.g., binding affinity, binding rate, kinetics, etc.

Microfluidic devices are capable of automation. With or without external pumping system/equipment/tools, microfluidics is capable of integrating sample pre- and post-processing, sensing, surface modification, temperature control, EM field control, etc. The automation of microfluidic-integrated sensors and structures is reviewed by [30, 32–34]. However, the automation of whole microfluidic-integrated biosensor as one device still seems quite challenging as the liquid handling in this field is usually more complex which could involve up to dozens of solutions and operations like filtering, centrifugation, etc., together with the activity of biological samples to be considered. Partially automized microfluidic-integrated DNA biosensors are reviewed by Ansari [35]. Joung [36] presented a novel lateral flow immunosensor (LFI) for microfluidic-integrated enzyme immunosorbent assay (EIA) in POC testing (POCT), a chemiluminescent LFI-based automatic EIA system, the operation of which does not require additional steps such as mechanical fluidic control, washing, or injecting. The key concept relies on a delayed-release effect of chemiluminescent substrates (luminol enhancer and hydrogen peroxide generator) by an asymmetric polysulfone membrane (ASPM). When the ASPM was placed between the nitrocellulose membrane and the substrate pad, substrates encapsulated in the substrate pad were released after 5.3 ± 0.3 min. As a proof of concept, the high-sensitivity C-reactive protein level in human serum was detected by this sensor.

Microfluidics enables both separate and mutual processing of multiple binding assays with single or multiple samples simultaneously. For the detection of single or multiple targets in a real or complex solution, the design of the binding assay usually involves more than one kind of bioreceptors, thus meaning that the designed samples to flow over each bioreceptor spot could be different. The delivering of different kinds of samples in sequential orders can be realized by unique design of microfluidic channels, pneumatic valves [17, 37], and/or centrifugal forces [33, 34].

Microfluidic structures ensure the precise control over experimental conditions [38]. What can be precisely controlled by microfluidic structures include flow rate, sample volume, channel volume, channel height, reaction time, etc. Integration of sensors with microfluidic channels serves to reduce assay time by constraining the diffusion distance between the molecules in the sample and the sensor and to create laminar flow over the sensor to distribute target molecules broadly and uniformly [2].

5. The present of microfluidic-integrated biosensors

Biosensors can be classified based on target recognition events and transduction mechanisms [4]. Based on the target recognition events, biosensor receptors are included. Based on the transduction mechanisms, biosensors can be classified into optical biosensor (Raman scattering [39–44], surface plasmon resonance (SPR) [6, 45–47], fiber Bragg grating [48–53], fluorescent [54–58], chemiluminescence [36, 59]), electrochemical biosensor [60–64], calorimetric biosensor [6, 22, 65–69], and piezoelectric biosensor [70–74].

5.1 Target recognition

Biological targets to be detected by biosensors, especially for the detection of analytes holden by human beings/animals, could be divided into two kinds, i.e., physical parameters and physiological/biological targets. Physical parameters like the body temperature, blood pressure, heart rate, velocity, and location usually do not request a corresponding and unique bioreceptor on the biosensor, as these physical parameters usually can be detected directly by optical, electronic, and piezoelectric sensors. Analytes as physiological/biological targets, however, cannot be detected directly, because of the complex components in a real human sample, so bioreceptors are adopted for the specific recognition of these targets, including cell, antibodies, DNAs, aptamers, and molecularly imprinted polymers [4].

The most commonly adopted physiological fluids of human beings/animals are blood, which has to be collected in an invasive way, and fluids that can be collected in a noninvasive way, e.g., sweat, saliva, tears, and urine, can be used in the prediction and diagnosis of various diseases [75–77]. Comparing with other physiological fluids, saliva is the outstanding fluid with the advantages of easy accessing and large volume, but with a major disadvantage of large range of variability in components and concentrations depending on the extent of oral cleanliness; examples that have been experimentally verified are using human saliva for the detection of cytokine [78], dopamine [51], insulin [79], fetuin [80], bacterial load [81], cholesterol [25], and cortisol [82]; using tear for the detection of dopamine [83], proteomic, lipidomic, and metabolomic composition [77]; using sweat for the detection of cytokine [84] and proteomic [76]; and using urine for the detection of anticancer drugs [85], L-carnitine [86], *Chlamydia trachomatis*, and *Neisseria gonorrhoeae* [87]. Samples of sweat and tear have been significantly undeveloped until quite recent when flexible materials and flexible electronic techniques achieved some milestones [4]. Currently the most well-explored targets in human physiological fluids include electrolytes (e.g., K^+, Ca^{2+}, Na^+) and major metabolites (e.g., myocardial enzyme, glucose, urea), which lack specification to diseases, indicating the general physiological conditions [4].

5.2 Transduction mechanism

So far, optical biosensors deliver the best sensitivity among the three other kinds of biosensors; electrochemical biosensors are the most popular choice as commercial-potential biosensors because of the compact size, low cost, and acceptable sensitivity; colorimetric biosensors are with a distinguished advantage of easy operation at extremely low cost but with a major disadvantage of low sensitivity; while the researches on piezoelectric biosensors are quite limited comparing with three other kinds of biosensors. Some most up-to-date researches on all fours kinds of biosensors are presented below.

5.2.1 Optical biosensors

Surface-enhanced Raman spectroscopy (SERS) and surface plasmon resonance are the two powerful optical biosensors with a unique feature of label-free sensing, as the analytes need no pre-processing to be labeled before sensing events and thus eliminate the false-positive or false-negative biosensing results induced by the labels. The first commercial product of SPR biosensor appeared in 1990 by the company of Pharmacia (named Biacore afterwards). Since then, more than 1000 papers were published annually using commercial SPR biosensors [88]. Most of these commercial SPR biosensors are bulky and only laboratory based. The development of plasmonic-based biosensors in the field of POC was reviewed in [43] together with recent advances in surface chemistry, substrate fabrication, and microfluidic integration. Here we explore a bit wider which is not limited to POC but microfluidic-integrated biosensors. In most of the researches mentioned below, the microfluidic structure usually serves as the sample handling unit.

Tunc I et al. [39] presented the molecular specificity of Raman spectroscopy together with self-assembled monolayer of metallic AuNPs to detect CA125 antibody–antigen molecules. Highly enhanced electromagnetic fields localized around neighboring AuNPs provide hot-spot construction due to the spatial distribution of SERS enhancement on the CA125 proteins at nM concentration level.

Carneiro M et al. [41] reported the detection of carcinoembryonic antigen (CEA) in SERS using two different bioreceptors for CEA, i.e., a molecularly imprinted polymer (MIP) and a natural antibody. The MIP acted as a pre-concentration scheme for the CEA, while the natural antibody signals the presence of CEA on the MIP platform. The MIP film was first incubated in the sample containing CEA and next incubated in SERS tag, which is gold nano-stars coupled to 4-aminothiophenol (4-ATP) as Raman reporter, so the MIP acted as a pre-concentration scheme for the CEA. Then the MIP was exposed to the natural CAE antibody. A sensitivity down to 1.0 ng/mL was reported.

Zhu JY et al. [89] presented a biosensor that can be used for clinical diagnosis. This biosensor is based on localized surface plasmon resonance integrated with a biomimetic microfluidic "adipose-tissue-on-chip" platform for an in situ label-free, high-throughput, and multiplexed cytokine secretion analysis of obese adipose tissue. It was stated that this system enables simultaneous measurements of pro-inflammatory (IL-6 and TNF-alpha) and anti-inflammatory (IL-10 and IL-4) cytokines secreted by the adipocytes and macrophages and identified stage-specific cytokine secretion profiles from a complex milieu during obesity progression.

In the research of [90], the plasmonic biosensor integrated the microfluidic unit for plasma separation, allows the in-line separation of plasma directly from the bloodstream without any pre-processing outside the device, and channels it to the active detection area, where inorganic cerium oxide nanoparticles function as local selective dopamine binding sites through strong surface redox reaction. A detection limit of dopamine was achieved at 100 fM concentration in simulated body fluid and 1 nM directly from blood without any prior sample preparation. This demonstration shows the feasibility of the practical implementation of the proposed plasmonic system in detection of a variety of biomarkers directly from the complex biological fluids. Li XK et al. [91] reported the plasmonic biosensor integrated a multifunctional microfluidic system with small-volume microchamber and regulation channels for reliable monitoring of cytokine secretion from individual cells for hours.

Besides the traditional plasmonic materials, graphene has recently received more and more attention in the field of both labeled and label-free sensing, because of its ability to harness electromagnetic fields, strong light-matter interaction of graphene layer, and its highly tunable optical properties [40]. Liu HP et al. [40] simulated the detection capacity of the graphene plasmonic biosensor using three-dimensional finite difference time domain method. Numerical results showed that the maximum sensitivity and figure of merit of the biosensor are 333.3 nm/RIU and 16.665 RIU, respectively.

Fluorescence is the other powerful optical biosensor which labels analytes and promises high sensitivity and specificity in target recognition. Raducanu VS et al. [56] reported a direct fluorescent signal transducer embedded in a DNA aptamer for versatile metal-ion detection. This sensor embedded with guanine-rich DNA aptamer internally coupled with Cy3 fluorescent dye that measures directly the DNA conformational changes upon metal-ion binding. Our signal transducer is environmentally sensitive that is internally coupled to the DNA aptamer. Potassium ion concentration was successfully measured in a variety of aqueous and biological test samples.

5.2.2 Electrochemical biosensors

There are plenty of researches on electrochemical biosensors, and majority of the commercialized biosensors belong to this category. Here we only present the electrochemical biosensors integrated with microfluidics that possesses both miniaturized structure and high sensitivity.

Electrode-based chemoelectrical biosensors are the most common ones. Usually a working electrode and a blank/reference electrode are designed in such biosensor, and the samples cover both electrodes and generate a measurable electrical signal. Mi SL et al. [92] reported a sensitivity up to 567 nA mM(−1) mm(−2), and the limit of detection was 4.5 M (vs. Ag/AgCl as the reference electrode) in the detection of metabolic lactate concentrations in HepG2 cells cultured with cancer drugs.

Evans D et al. [93] demonstrated a fully integrated microfluidic amperometric enzyme-linked immunosorbent assay prototype using a commercial interferon gamma release assay as a model antibody binding system. What is unique in this research is that the assay cell is based on a printed circuit board (PCB) and the microfluidic assay chemistry was engineered to take place on the Au-plated electrodes within the cell. All components were manufactured exclusively via standard commercial PCB fabrication processes. A detection limit comparable to high-end commercial systems and a short diagnosis time of 8 minutes were demonstrated.

Silicon nanowire field-effect transistor is one of the most sensitive biosensing techniques, but it is limited to analytes that carry charges. Weakly charged or uncharged analytes can hardly be detected [11]. Evans D et al. [31] presented a method of immobilizing bioreceptors on the silicon nanowire sensing surface only, comparing with the traditional methods in which a large surrounding substrate is also covered with bioreceptors, and it was proven that restricting the surface modification substantially improves the sensitivity.

Besides silicon nanowire, copper nanowire is adopted in electrochemical biosensors [94]. In [94], microfluidic chip is coupled to copper nanowires for the fast diagnosis of galactosemia in precious newborn urine samples. Galactosemia is a rare disease that is diagnosed through the identification of different metabolite profiles. The specific detection of galactose 1-phosphate (Gal 1-P), galactose (Gal), and uridyl diphosphate galactose (UDP-Gal) confirms type I, II, and III galactosemia diseases. The detection is extremely fast which is less than 350 s, required negligible urine sample consumption, and displayed impressive signal-to-noise characteristics and excellent reproducibility.

Oliveira MC et al. [95] presented an amperometric biosensor using a screen-printed electrode modified with carbon nanotubes and nickel ions for the detection of glucose, which is characterized by the chemical oxidation of carbohydrate by NiOOH. Under optimized conditions, a limit of detection 3.9 μmol/L and a limit of quantification of 13 μmol/L were reported. The effect of concomitant species such as ascorbic acid, dopamine, and uric acid was investigated, and this method was successfully applied for the determination of glucose in a commercial blood serum human (original and spiked) sample. What is unique in this research is that the microfluidic system was assembled on a 3D-printed platform constructed with acrylonitrile butadiene styrene and integrated with nine cotton threads, providing a stable flow rate.

5.2.3 Colorimetric biosensors

Plenty of reports are available on colorimetric biosensors integrated with microfluidics, e.g., [27, 28, 96–98]; most of the reports highlighted the features of cost-effectiveness and miniaturization. Different from three other kinds of biosensors, the materials adopted for the integrated microfluidic structures are usually not PDMS, but paper for the majority and cloth in some researches. Currently majority of the researches focus on the applications in food safety [22, 27, 96] and heavy metal detection [67, 99, 100] in aqueous environment. The researches in the application of biological analytes are quite limited, due to the natures of analytes and bioreceptors and the environment conditions in order to keep the activities of both analytes and bioreceptors.

Fraser LA et al. [101] presented a malaria biosensor whereby aptamers are coated onto magnetic microbeads for magnet-guided capture, wash, and detection of the biomarker. A biosensor incorporating three separate microfluidic chambers was designed to enable such magnet-guided equipment-free colorimetric detection of PfLDH. The biosensor showed high sensitivity and specificity when detecting PfLDH using both in vitro cultured parasite samples and clinical samples from malaria patients.

5.2.4 Piezoelectric biosensors

The research of piezoelectric biosensor integrated with microfluidics is quite underdeveloped so far. Possible reasons could be the lower sensitivity, poor biocompatibility, and complicated fabrication.

Yamaguchi M. [82] proposed a mass sensor based on mechanical resonance that incorporates a disk-shaped mechanical resonator, a separate piezoelectric element used to excite vibrations in the resonator, and a microfluidic mechanism. Electrical power is used to actuate the piezoelectric element, leaving the resonator free from power lines. This sensor was reported to be suitable to analyze the concentration of a salivary hormone, cortisol in human saliva samples.

6. Future research interests

Future possible research interests in the field of microfluidic-integrated biosensors are proposed as the following:

- Exploration of materials for both microfluidics and nanofluidics in different application scenarios. Besides PDMS, the exploration of other materials, e.g., engineering polymers, traditional glass, silicon, or metal, in special applications that requires high chemical stability, high thermal stability, unique optical properties, and/or special mechanical properties.

- Fabrication of microfluidics and nanofluidics and structures (e.g., valves, mixers).

- Microfluidics with high chemical and thermal stability for special applications.

- Integration of microfluidics and nanofluidics with sensors to form complete and functional systems that require no professional operations and ease in applications, e.g., LOC, etc.

- Integration of microfluidics with data processing algorithms. The application of machine learning in sensing data processing could enhance the performance of biosensors in specialized environments.

- Integration of microfluidics with communication techniques. Synchronization of sensing data with relevant users, remote control of the biosensors, and big data analysis of special sensing networks can be realized.

7. Conclusions

The state-of-the-art advances in biosensor development based on microfluidic technology have been reviewed in the book chapter with focus on the applications, challenges, and possible future research interests for each type of biosensor. It can be envisioned that microfluidic-based biosensors will remain a hot topic of investigations because of the ever-increasing demands in various applications ranging from industry to biomedical detection. The interests in microfluidic-integrated biosensors promise even more prospective future in these areas. It is applications in wearable biosensor; portable biosensor can be explored in the future with enhanced sensitivity, improved stability, and miniaturized structure.

Acknowledgements

This research was supported by Zhejiang Provincial Natural Science Foundation of China under Grant No. LQ19F050003 as well as Ningbo Science and Technology Bureau under Service Industry Science & Technology Programme with project code 2019F1030.

Conflict of interest

The authors declare no conflict of interest.

Author details

Jing Wang[1*], Yong Ren[2] and Bei Zhang[3*]

1 Department of Electrical and Electronic Engineering, University of Nottingham Ningbo China, Zhejiang, China

2 Department of Mechanical, Materials and Manufacturing Engineering, University of Nottingham Ningbo China, Zhejiang, China

3 Department of Automation Science and Electrical Engineering, Beihang University, Beijing, China

*Address all correspondence to: jing.wang@nottingham.edu.cn; bei.zhang@buaa.edu.cn

References

[1] Schasfoort RBM, Tudos AJ. Handbook of Surface Plasmon Resonance. Royal Soc Chemistry; 2008

[2] Che C et al. Activate capture and digital counting (AC + DC) assay for protein biomarker detection integrated with a self-powered microfluidic cartridge. Lab on a Chip. 2019;**19**(23):3943-3953

[3] Vashist SK et al. Emerging technologies for next-generation point-of-care testing. Trends in Biotechnology. 2015;**33**(11):692-705

[4] Tu J et al. The Era of Digital Health: A Review of Portable and Wearable Affinity Biosensors. Advanced Functional Materials. 2019:1906713

[5] Steinhubl SR, Muse ED, Topol EJ. The emerging field of mobile health. Science Translational Medicine. 2015;**7**(283):283rv3

[6] Wang J et al. Highly sensitive multipoint real-time kinetic detection of surface plasmon bioanalytes with custom CMOS cameras. Biosensors and Bioelectronics. 2014;**58**:157-164

[7] Ciampi S, Harper JB, Gooding JJ. Wet chemical routes to the assembly of organic monolayers on silicon surfaces via the formation of Si–C bonds: Surface preparation, passivation and functionalization. Chemical Society Reviews. 2010;**39**(6):2158-2183

[8] Liang G et al. Fiber Optic Surface Plasmon Resonance-Based Biosensor Technique: Fabrication, Advancement, and Application. Critical Reviews in Analytical Chemistry. 2016:213-223

[9] Mok J et al. Digital microfluidic assay for protein detection. Proceedings of the National Academy of Sciences. 2014;**111**(6):2110-2115

[10] Sun J, Wang S, Gao F. Covalent surface functionalization of semiconducting polymer dots with beta-cyclodextrin for fluorescent Ratiometric assay of cholesterol through host-guest inclusion and FRET. Langmuir. 2016;**32**(48):12725-12731

[11] Shen M-Y, Li B-R, Li Y-K. Silicon nanowire field-effect-transistor based biosensors: From sensitive to ultra-sensitive. Biosensors and Bioelectronics. 2014;**60**:101-111

[12] Chu CJ et al. Improving nanowire sensing capability by electrical field alignment of surface probing molecules. Nano Letters. 2013;**13**(6):2564-2569

[13] Schneider CS et al. Surface plasmon resonance as a high throughput method to evaluate specific and non-specific binding of nanotherapeutics. Journal of Controlled Release. 2015;**219**:331-344

[14] Wuethrich A, Howard CB, Trau M. Geometric optimisation of electrohydrodynamic fluid flows for enhanced biosensing. Microchemical Journal. 2018;**137**:231-237

[15] Arreola J et al. Surface functionalization for spore-based biosensors with organosilanes. Electrochimica Acta. 2017;**241**:237-243

[16] Platt GW et al. Allergen immobilisation and signal amplification by quantum dots for use in a biosensor assay of IgE in serum. Biosensors and Bioelectronics. 2014;**52**:82-88

[17] Whitesides GM. The origins and the future of microfluidics. Nature. 2006;**442**(7101):368-373

[18] Mross S et al. Microfluidic enzymatic biosensing systems: A review. Biosensors and Bioelectronics. 2015;**70**:376-391

[19] Yetisen AK, Akram MS, Lowe CR. Paper-based microfluidic point-of-care diagnostic devices. Lab on a Chip. 2013;**13**(12):2210-2251

[20] Li X, Ballerini DR, Shen W. A perspective on paper-based microfluidics: Current status and future trends. Biomicrofluidics. 2012;**6**(1):011301

[21] Nie Z et al. Integration of paper-based microfluidic devices with commercial electrochemical readers. Lab on a Chip. 2010;**10**(22):3163-3169

[22] Aksorn J, Teepoo S. Development of the simultaneous colorimetric enzymatic detection of sucrose, fructose and glucose using a microfluidic paper-based analytical device. Talanta. 2020;**207**:120302

[23] Chethana K et al. Fiber bragg grating sensor based device for simultaneous measurement of respiratory and cardiac activities. Journal of Biophotonics. 2017;**10**(2):278-285

[24] Hung TQ et al. A novel lab-on-chip platform with integrated solid phase PCR and supercritical angle fluorescence (SAF) microlens array for highly sensitive and multiplexed pathogen detection. Biosensors and Bioelectronics. 2017;**90**:217-223

[25] Lee YJ et al. Enzyme-loaded paper combined impedimetric sensor for the determination of the low-level of cholesterol in saliva. Sensors and Actuators B: Chemical. 2018;**271**:73-81

[26] López Marzo AM et al. All-integrated and highly sensitive paper based device with sample treatment platform for Cd2+ immunodetection in drinking/tap waters. Analytical Chemistry. 2013;**85**(7):3532-3538

[27] Mooltongchun M, Teepoo S. A simple and cost-effective microfluidic paper-based biosensor analytical device and its application for hypoxanthine detection in meat samples. Food Analytical Methods. 2019;**12**(12):2690-2698

[28] Rafatmah E, Hemmateenejad B. Colorimetric and visual determination of hydrogen peroxide and glucose by applying paper-based closed bipolar electrochemistry. Microchimica Acta. 2019;**186**(11)

[29] Guan W, Liu M, Zhang CS. Electrochemiluminescence detection in microfluidic cloth-based analytical devices. Biosensors & Bioelectronics. 2016;**75**:247-253

[30] Alam MK et al. Recent advances in microfluidic technology for manipulation and analysis of biological cells (2007-2017). Analytica Chimica Acta. 2018;**1044**:29-65

[31] Li B-R et al. Biomolecular recognition with a sensitivity-enhanced nanowire transistor biosensor. Biosensors and Bioelectronics. 2013;**45**:252-259

[32] Maceiczyk RM, Lignos IG, DeMello AJ. Online detection and automation methods in microfluidic nanomaterial synthesis. Current Opinion in Chemical Engineering. 2015;**8**:29-35

[33] Burger R, Amato L, Boisen A. Detection methods for centrifugal microfluidic platforms. Biosensors and Bioelectronics. 2016;**76**:54-67

[34] Torres Delgado SM et al. Wirelessly powered and remotely controlled valve-array for highly multiplexed analytical assay automation on a centrifugal microfluidic platform. Biosensors and Bioelectronics. 2018;**109**:214-223

[35] Ansari MIH et al. Microfluidic-integrated DNA nanobiosensors.

Biosensors and Bioelectronics. 2016;**85**:247-260

[36] Joung H-A, Oh YK, Kim M-G. An automatic enzyme immunoassay based on a chemiluminescent lateral flow immunosensor. Biosensors and Bioelectronics. 2014;**53**:330-335

[37] Pradeep A et al. Automated and programmable electromagnetically actuated valves for microfluidic applications. Sensors and Actuators a-Physical. 2018;**283**:79-86

[38] Halldorsson S et al. Advantages and challenges of microfluidic cell culture in polydimethylsiloxane devices. Biosensors and Bioelectronics. 2015;**63**:218-231

[39] Tunc I, Susapto HH. Label-free detection of ovarian Cancer antigen CA125 by surface enhanced Raman scattering. Journal of Nanoscience and Nanotechnology. 2020;**20**(3):1358-1365

[40] Liu HP et al. A 3D multilayer curved plasmonic coupling array with abundant and uniform hot spots for surface-enhanced Raman scattering. Journal of Physics D: Applied Physics. 2020:53(5)

[41] Carneiro M et al. Dual biorecognition by combining molecularly-imprinted polymer and antibody in SERS detection. Application to carcinoembryonic antigen. Biosensors & Bioelectronics. 2019:146

[42] Lee T et al. Single functionalized pRNA/gold nanoparticle for ultrasensitive MicroRNA detection using electrochemical surface-enhanced Raman spectroscopy. Advanced Science.

[43] Li ZH et al. Plasmonic-based platforms for diagnosis of infectious diseases at the point-of-care. Biotechnology Advances. 2019;**37**(8):107440

[44] Sakir M et al. Photocatalytic green fabrication of Au nanoparticles on ZnO nanorods modified membrane as flexible and photocatalytic active reusable SERS substrates. Colloids and Surfaces a-Physicochemical and Engineering Aspects. 2020;**585**

[45] Bhardwaj H, Sumana G, Marquette CA. A label-free ultrasensitive microfluidic surface Plasmon resonance biosensor for Aflatoxin B1 detection using nanoparticles integrated gold chip. Food Chemistry. 2020;**307**:125530

[46] Das CM et al. A theoretical insight into the use of anti-reflective coatings for the upliftment of sensitivity of surface plasmon resonance sensors. Optics Communications. 2020;**458**

[47] Farmani H, Farmani A, Biglari Z. A label-free graphene-based nanosensor using surface plasmon resonance for biomaterials detection. Physica E-Low-Dimensional Systems & Nanostructures. 2020;**116**

[48] Chen XY et al. In-situ detection of small biomolecule interactions using a plasmonic tilted fiber grating sensor. Journal of Lightwave Technology. 2019;**37**(11):2792-2799

[49] He YL et al. Fiber brag grating monitoring of a morphing wing based on a polyvinyl chloride reinforced silicone substrate. Optical Fiber Technology. 2019;**50**:145-153

[50] Lao JJ et al. Gold nanoparticle-functionalized surface Plasmon resonance optical Fiber biosensor: In situ detection of thrombin with 1 n.M detection limit. Journal of Lightwave Technology. 2019;**37**(11):2748-2755

[51] Lobry M et al. Non-enzymatic D-glucose plasmonic optical fiber grating biosensor. Biosensors & Bioelectronics. 2019;**142**

[52] Loyez M et al. Functionalized gold electroless-plated optical fiber gratings for reliable surface biosensing. Sensors and Actuators B: Chemical. 2019;**280**:54-61

[53] Sypabekova M et al. Functionalized etched tilted fiber Bragg grating aptasensor for label-free protein detection. Biosensors & Bioelectronics. 2019;**146**

[54] Fan XC et al. Target-induced autonomous synthesis of G-quadruplex sequences for label-free and amplified fluorescent aptasensing of mucin 1. Sensors and Actuators B: Chemical. 2020;**304**

[55] Li J et al. Glucose assay based on a fluorescent multi-hydroxyl carbon dots reversible assembly with phenylboronic acid brush grafted magnetic nanoparticles. Sensors and Actuators B: Chemical. 2020;**304**

[56] Raducanu VS et al. A direct fluorescent signal transducer embedded in a DNA aptamer paves the way for versatile metal-ion detection. Sensors and Actuators B: Chemical. 2020;**304**

[57] Tang YY et al. A robust OFF-ON fluorescent biosensor for detection and clearance of bacterial endotoxin by specific peptide based aggregation induced emission. Sensors and Actuators B: Chemical. 2020;**304**

[58] Zhang JY et al. F-containing initiatior for ultrasensitive fluorescent detection of lung cancer DNA via atom transfer radical polymerization. Analytica Chimica Acta. 2020;**1094**:99-105

[59] Yu J et al. Microfluidic paper-based chemiluminescence biosensor for simultaneous determination of glucose and uric acid. Lab on a Chip. 2011;**11**(7):1286-1291

[60] Hu HL et al. Ni hierarchical structures supported on Titania nanowire arrays as efficient nonenzymatic glucose sensor. Journal of Nanoscience and Nanotechnology. 2020;**20**(5):3246-3251

[61] Hui YY et al. An electrochemical aptasensor based on DNA-AuNPs-HRP nanoprobes and exonuclease-assisted signal amplification for detection of aflatoxin B-1. Food Control. 2020;**109**

[62] Peyman H, Roshanfekr H, Ansari S. DNA-based electrochemical biosensor using chitosan-carbon nanotubes composite film for biodetection of Pirazon. Eurasian Chemical Communication. 2020;**2**(2):213-225

[63] Sundar S, Kwon SJ, Venkatachalam G. Magneto-biosensor for the detection of uric acid using citric acid-capped Iron oxide nanoparticles. Journal of Nanoscience and Nanotechnology. 2020;**20**(4):2144-2153

[64] Vathani AM et al. Fabrication of Al-TiO2 thin film electrode by spray pyrolysis technique for urea sensing. Journal of Nanoscience and Nanotechnology. 2020;**20**(5):2887-2892

[65] Fu QQ et al. Ambient light sensor based colorimetric dipstick reader for rapid monitoring organophosphate pesticides on a smart phone. Analytica Chimica Acta. 2019;**1092**:126-131

[66] Huang LT, Li ZH, Guo LQ. Colorimetric assay of acetylcholinesterase inhibitor tacrine based on MoO2 nanoparticles as peroxidase mimetics. Spectrochimica Acta Part a-Molecular and Biomolecular Spectroscopy. 2020;**224**

[67] Jia M et al. Extended GR-5 DNAzyme-based autonomous isothermal Cascade machine: An efficient and sensitive

one-tube colorimetric platform for Pb2+ detection. Sensors and Actuators B: Chemical. 2020;**304**

[68] Kong LB et al. A novel smartphone-based CD-spectrometer for high sensitive and cost-effective colorimetric detection of ascorbic acid. Analytica Chimica Acta. 2020;**1093**:150-159

[69] Wei SQ et al. Exponential amplification reaction and triplex DNA mediated aggregation of gold nanoparticles for sensitive colorimetric detection of microRNA. Analytica Chimica Acta. 2020;**1095**:179-184

[70] Kim KI et al. Influence of temperature and humidity on the detection of benzene vapor by a piezoelectric crystal sensor. Instrumentation Science & Technology. 2019;**47**(4):436-447

[71] Pohanka M. Piezoelectric Immunosensor for the determination of C-reactive protein. International Journal of Electrochemical Science. 2019;**14**(9):8470-8478

[72] Urdinola KB et al. In-Silico Prediction on the MSAMS-Assisted Immobilization of Bovine Serum Albumin on 10 MHz Piezoelectric Immunosensors. Journal of Molecular and Engineering Materials. 2019;**7**(1-2)

[73] Yuan M et al. Piezoelectric arsenite aptasensor based on the use of a self-assembled mercaptoethylamine monolayer and gold nanoparticles. Microchimica Acta. 2019;**186**(5)

[74] Zamora-Sequeira R et al. What are the Main Sensor Methods for Quantifying Pesticides in Agricultural Activities? A Review. Molecules. 2019:24(14)

[75] Katchman BA et al. Eccrine Sweat as a Biofluid for Profiling Immune Biomarkers. Proteomics Clinical Applications. 2018:12(6)

[76] Adewole OO et al. Proteomic profiling of eccrine sweat reveals its potential as a diagnostic biofluid for active tuberculosis. Proteomics Clinical Applications. 2016;**10**(5):547-553

[77] Hagan S, Martin E, Enriquez-de-Salamanca A. Tear fluid biomarkers in ocular and systemic disease: Potential use for predictive, preventive and personalised medicine. Epma Journal. 2016;**7**

[78] Belstrøm D et al. Salivary cytokine levels in early gingival inflammation. Journal of Oral Microbiology. 2017;**9**(1)

[79] Viswanath B et al. Recent trends in the development of diagnostic tools for diabetes mellitus using patient saliva. Trends in Analytical Chemistry. 2017;**89**:60-67

[80] Sánchez-Tirado E et al. Magnetic multiwalled carbon nanotubes as nanocarrier tags for sensitive determination of fetuin in saliva. Biosensors and Bioelectronics. 2018;**113**:88-94

[81] Ana C et al. Dental caries and bacterial load in saliva and dental biofilm of type 1 diabetics on continuous subcutaneous insulin infusion. Journal of Applied Oral Science. 2018;**26**

[82] Yamaguchi M. Microfluidic line-free mass sensor based on an antibody-modified mechanical resonator. Micromachines. 2018;**9**(4):13

[83] Sharma N et al. Dopamine levels in human tear fluid. Indian Journal of Ophthalmology. 2019;**67**(1):38-41

[84] Hladek MD et al. Using sweat to measure cytokines in older adults compared to younger adults: A pilot study. Journal of Immunological Methods. 2018;**454**:1-5

[85] Chandra P et al. Separation and simultaneous detection of anticancer

drugs in a microfluidic device with an amperometric biosensor. Biosensors and Bioelectronics. 2011;**28**(1):326-332

[86] Andrianova MS et al. CMOS-compatible biosensor for L-carnitine detection. Biosensors and Bioelectronics. 2018;**119**:48-54

[87] Soler M et al. Multiplexed nanoplasmonic biosensor for one-step simultaneous detection of chlamydia trachomatis and Neisseria gonorrhoeae in urine. Biosensors and Bioelectronics. 2017;**94**:560-567

[88] Myszka DG. Handbook of surface Plasmon resonance 2nd edition foreword to the 1st edition. In: Schasfoort RBM, editor. Handbook of surface Plasmon resonance, 2nd edition. 2017. pp. V-IX

[89] Zhu JY et al. An integrated adipose-tissue-on-chip nanoplasmonic biosensing platform for investigating obesity-associated inflammation. Lab on a Chip. 2018;**18**(23):3550-3560

[90] Vazquez-Guardado A et al. Enzyme-free plasmonic biosensor for direct detection of neurotransmitter dopamine from whole blood. Nano Letters. 2019;**19**(1):449-454

[91] Li XK et al. Label-Free Optofluidic Nanobiosensor Enables Real-Time Analysis of Single-Cell Cytokine Secretion. Small. 2018:14(26)

[92] Mi SL et al. An integrated microchannel biosensor platform to analyse low density lactate metabolism in HepG2 cells in vitro. RSC Advances. 2019;**9**(16):9006-9013

[93] Evans D et al. A novel microfluidic point-of-care biosensor system on printed circuit board for cytokine detection. Sensors. 2018;**18**(11)

[94] Garcia M, Alonso-Fernandez J, Escarpa A. Copper nanowires immobilized on the boards of microfluidic chips for the rapid and simultaneous diagnosis of Galactosemia diseases in Newborn urine samples. Analytical Chemistry. 2013;**85**(19):9116-9125

[95] Oliveira MC et al. Nonenzymatic sensor for determination of glucose in blood plasma based on nickel oxyhydroxide in a microfluidic system of cotton thread. Journal of Electroanalytical Chemistry. 2019;**840**:153-159

[96] Chungchai W et al. Development of a novel three-dimensional microfluidic paper-based analytical device (3D-mu PAD) for chlorpyrifos detection using graphene quantum-dot capped gold nanocomposite for colorimetric assay. International Journal of Environmental Analytical Chemistry

[97] Fakhri N et al. Paper based colorimetric detection of miRNA-21 using Ag/Pt nanoclusters. Spectrochimica Acta Part a-Molecular and Biomolecular Spectroscopy. 2020;**227**

[98] Li F et al. High-resolution temporally resolved chemiluminescence based on double-layered 3D microfluidic paper-based device for multiplexed analysis. Biosensors & Bioelectronics. 2019;**141**

[99] Hossain SMZ, Brennan JD. Beta-Galactosidase-based colorimetric paper sensor for determination of heavy metals. Analytical Chemistry. 2011;**83**(22):8772-8778

[100] Lin Y et al. Detection of heavy metal by paper-based microfluidics. Biosensors and Bioelectronics. 2016;**83**:256-266

[101] Fraser LA et al. A portable microfluidic Aptamer-tethered enzyme capture (APTEC) biosensor for malaria diagnosis. Biosensors and Bioelectronics. 2018;**100**:591-596

Chapter 5

Biological Bone Micro Grinding Temperature Field under Nanoparticle Jet Mist Cooling

Min Yang, Changhe Li, Liang Luo, Lan Dong, Dongzhou Jia, Runze Li, Mingzheng Liu, Xin Cui, Yali Hou, Yanbin Zhang, Teng Gao, Xiaoming Wang and Yunze Long

Abstract

Clinical neurosurgeons used micro grinding to remove bone tissues, and drip irrigation-type normal saline (NS) is used with low cooling efficiency. Osteonecrosis and irreversible thermal neural injury caused by excessively high grinding temperature are bottleneck problems in neurosurgery and have severely restricted the application of micro grinding in surgical procedures. Therefore, a nanoparticle jet mist cooling (NJMC) bio-bone micro grinding process is put forward in this chapter. The nanofluid convective heat transfer mechanism in the micro grinding zone is investigated, and heat transfer enhancement mechanism of solid nanoparticles and heat distribution mechanism in the micro grinding zone are revealed. On this basis, a temperature field model of NJMC bio-bone micro grinding is established. An experimental platform of NJMC bio-bone micro grinding is constructed, and bone micro grinding force and temperatures at different measuring points on the bone surface are measured. The results indicated that the model error of temperature field is 6.7%, theoretical analysis basically accorded with experimental results, thus certifying the correctness of the dynamic temperature field in NJMC bio-bone micro grinding.

Keywords: grinding, nanoparticle jet, mist cooling, biological bone, thermal injury, temperature field

1. Introduction

As a precise material removal method, grinding has been extensively applied to bone tissue removal and surface treatment in surgical procedures by conforming to the combining trend of machine science and biomedical science. In the machining field, the method of using miniature grinding tools (diameter is generally below 1 mm) for machining of feature size of 1~500 μm is called micro grinding [1]. As bone tissues processed in osteological surgery are viable tissues and diamond grinding tool has obviously higher specific energy in the grinding process than other cutting methods, human bone, nerves, and blood vessels can be easily influenced by high temperature, and when grinding temperature is higher than 50°C, bone tissues will experience irreversible necrosis, and nervous tissues will start experiencing thermal injury at 43°C [2]. In the present neurosurgery, surgeons usually perform drip irrigation of normal saline (NS) in the grinding zone following the following principle: The heat is carried away through convective heat transfer so as to cool the operative region. However, the drip irrigation-type cooling efficiency is not ideal; thermal injury in the operation process leads to necrosis of surrounding tissues and nerves, so the operation fails. The irreversible thermal injury of bone tissues triggered by high temperature in micro grinding is a bottleneck problem in bone micro grinding operation. Moreover, a large quantity of cooling liquid needs to be dripped into the grinding zone under drip irrigation-type cooling, which can easily reduce the visibility of the operative region under an endoscope [3].

In the machining field, as flood cooling grinding harms worker's health and the environment, dry grinding can easily result in serious burn of the specimen, and microdroplet jet grinding has insufficient cooling performance [4–7]; researchers have put forward nanoparticle jet mist cooling (NJMC) technique according to the theory of heat transfer enhancement, which cannot only enhance the convective heat transfer in the grinding zone by taking full advantages of high specific surface of nanoparticles and heat capacity but, meanwhile, can improve the tribological properties of the grinding zone and can reduce heat generation by virtue of superior antifriction and anti-wear characteristics and high bearing capacity of nanoparticles [8–10]. On this basis, as it is the fact that clinical surgical bone grinding has disadvantages of thermal injury and low visibility of operative region and that nanofluid microdroplet grinding has good cooling and lubrication effect with a small use level of cutting fluid in the machining field, the research team where the author joined proposed NJMC bio-bone micro grinding process: adding medical solid nanoparticles of a certain proportion into the NS and selecting the corresponding surface dispersant according to physical and chemical properties of nanoparticle material, supplemented by ultrasonic vibration, so medical nanoparticles can enjoy uniform and stable distribution in the NS so as to form a medical nanofluid under stable suspension, which is then sprayed into the focus grinding zone in jet form as carried by high-pressure gas via a miniature nozzle [11, 12]. And this is expected to lower the bone grinding temperature while improving the visibility of the operative region.

In recent years, researches have initially explored into thermal injury problem existing in the bio-bone grinding process. Taking heat transfer problem and its inverse problem in the bone grinding process with a miniature spherical grinding tool, Zhang et al. [13] conducted an in-depth study combining numerical simulation and experiment and analyzed the transient temperature field in the bone grinding process through a numerical simulation based on a heat generation model of bone grinding. Directing at the complex structure of bone issues and taking grinding temperature, grinding force, and blocking of grinding tool in the bone tissue grinding process as the study objects, Zhu et al. [14] obtained the grinding tool blocking mechanism in the bone tissue grinding by analyzing action rules of grinding temperatures on grinding temperature, grinding force, and grinding tool blocking. Sasaki et al. [15] analyzed heat generation during grinding using diamond tools with a diameter of 5 mm at a rotational speed of 100,000 r/min. They found that grinding for 10 cm along the grinding temperature without cooling liquid will expand the area of higher than 37.7°C continuously. To prevent thermal injuries to surrounding nerve tissues, Enomoto et al. [16] invented a new diamond spherical tool that can effectively reduce the temperature at the grinding zone. The surface of the diamond tool contains adsorbed super-hydrophilic TiO_2 nanoparticles, which effectively prevent temperature rise during bone grinding. The grinding zone is cooled during neurosurgery commonly by dripping a room-temperature NS into it.

The NS dissipates heat through natural heat convection, cooling the high-temperature region.

For the bottleneck problem in the current clinical neurosurgical bone operation, namely irreversible thermal injury, a NJMC bio-bone micro grinding process was put forward. The convective heat transfer mechanism of nanofluid in micro grinding zone was studied. A dynamic temperature field model in NJMC bio-bone grinding was constructed. By reference to the suppression measures of grinding thermal injury in the field of mechanical engineering, the bottleneck in orthopedic operation was solved so as to provide theoretical guidance and technical support for avoiding or reducing thermal injury in clinical orthopedic operation.

2. Grinding temperature field

Starting from the universal temperature field problem, the heat quantity Q at one spatial point takes place transiently, and the temperatures at other points nearby this point will be changed under the action of the transferred heat quantity, and meanwhile, they are variable due to the spatial–temporal change. Temperature field is a generic term of temperature distribution at spatial points at a certain time. Generally speaking, temperature field is a function of space and time [17]:

$$T = f(x, y, z, t) \tag{1}$$

Equation (1) denotes a three-dimensional (3D) unsteady-state temperature field where the object temperature is changed both in direction x, y, and z and at time t.

In this study, temperature field is divided into two types:

1. It is divided into steady-state temperature field and transient-state temperature field according to whether temperature changes with time. As indicated by Eq. (1), if the temperature field does not change with time, namely $\frac{\partial T}{\partial t} = 0$, it is a steady-state temperature field, or otherwise it is a transient-state temperature field. As mobile heat source is loaded, the temperature at specific point on the bone surface presents time-dependent change. Hence, the temperature field studied in this chapter is transient-state temperature field.

2. It is divided into temperature field in cut-in zone, that in steady-state zone and that in cut-out zone according to the grinding tool/specimen contact arc length. As shown in **Figure 1**, the effective cutting part of the grinding tool is totally within the length of the specimen material, the grinding tool/specimen contact arc length is l_c, and the temperature field is temperature field in steady-state zone. As the grinding process starts, the grinding tool/specimen

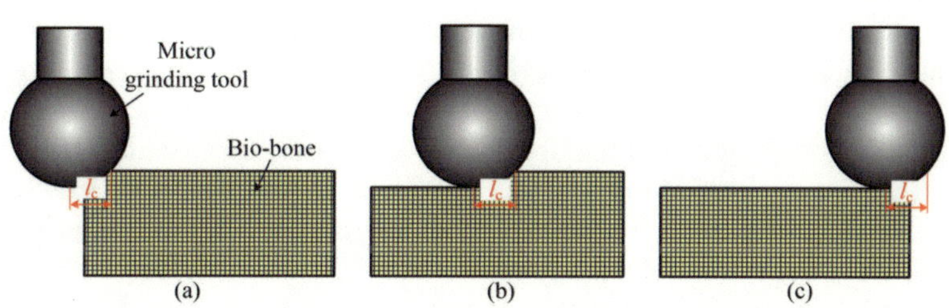

Figure 1.
Schematic diagram of abrasive/bone contact states. (a) Cut-in zone. (b) Steady-state zone. (c) Cut-out zone.

contact arc length is gradually enlarged, but not reaching l_c, and at the moment, the grinding tool/specimen material is in cut-in phase, so the temperature field is in cut-in zone. As the grinding tool starts moving out of the specimen length, the contact arc length is gradually reduced to 0, and at the time, the grinding tool cuts out the specimen material, so the temperature is in cut-out zone.

2.1 The solution method of grinding temperature field

The solving method of grinding temperature field includes analytical method and finite difference method.

2.1.1 Solution of grinding temperature field by analytical method

Following the heat transfer theory and energy conservation law, analytical method solves the temperature rise function based on all kinds of boundary conditions in actual grinding machining so as to obtain temperature values on specimen surface and each internal node. The advantage of the analytical method is that it cannot only obtain the functional relationship regarding temperature distribution but also analyze different influence factors related to temperature field and their influence laws on temperature field distribution. Based on the theory of mobile heat source proposed by Jaeger [18] in 1942, many researchers have established theoretical grinding heat models with the main idea being superposition method of heat source temperature field. In other words, grinding interface is regarded as surface heat source constituted by numerous linear heat sources while linear heat sources are treated as a combination of numerous tiny unit linear heat sources, and each tiny unit linear heat source is simplified into combined action of point heat sources. Therefore, the basis for superposition method of heat source temperature fields is the solution of this temperature field at any time after transient point heat sources in an infinite object instantaneously emit partial heat quantity.

2.1.2 Solution of grinding temperature field by finite difference method

As for solving the grinding temperature field, it can be complicated to use analytical method to solve even simple heat conduction problem. Grinding machining is itself of high complexity relative to other machining modes [19, 20], where input parameters of grinding temperature field are miscellaneous, abrasive particle distribution is irregular, grinding state of abrasive particle (plowing, sliding, cutting) is uncertain, cooling medium participates in convective heat transfer in the grinding zone, and surrounding airflow field of grinding tool has an effect on temperature. A large quantity of nonlinear coupling relations exists in the grinding process, so if any input parameter is changed, it will influence the follow-up derivation of expressions, and the solving of temperature field becomes even harder for the analytical method [21, 22]. Under this circumstance, the finite element method based on numerical method is a very effective method of solving heat conduction problem. It is only necessary to determine boundary conditions and initial conditions in order to conveniently calculate grinding temperature field, so it has been widely used by researchers at present. As the finite element method makes many hypotheses for grinding temperature field and boundary conditions, only specific built-in modules of specific software can be used [23], and the deviation of calculated grinding temperature from actual temperature is large. The finite element method based on numerical method is another effective method of calculating grinding temperature field between analytical method and finite element method.

It can accurately calculate temperature field through a theoretical modeling of boundary conditions of temperature field (heat flux, heat distribution ratio, convective heat transfer coefficient, etc.) according to actual grinding conditions. There have been few reports on calculation of grinding temperature field via finite difference method, so this method will be hereby described in details.

2.1.2.1 Basic principle of finite difference method

The object is divided into finite grid cells. A difference equation is obtained by transforming the differential equation, and the temperature at each grid cell node is solved through numerical simulation. As shown in **Figure 2**, under the 2D heat conduction problem, the object is divided into rectangular grids along directions x and y according to the spacing between Δx and Δy. The node is defined as intersection point of each grid line, $p\ (i, j)$ denotes the position of each node, i is serial number at node along the direction x, j is serial number at node along the direction y, and the intersection point between object boundary and grid is defined as boundary node. The basic principle of this method is to replace differential quotient with finite difference quotient so as to transform the original differential equation into a difference Equation [24].

2.1.2.2 Establishment of differential equation of heat conduction

In the grinding heat conduction problem, Fourier's law is the most fundamental heat conduction equation, namely the heat quantity passing through infinitesimal isothermal surface A within limited time interval t is Q, which is in direct proportion of temperature gradient $\frac{\partial T}{\partial n}$ and is contrary to the direction of temperature field:

$$dQ = -k_{\mathrm{w}} \frac{\partial T}{\partial n} dA dt \tag{2}$$

For heat flux:

$$q_{\mathrm{x}} = -k_{\mathrm{w}} \frac{\partial T}{\partial x} \tag{3}$$

where q_{x} is heat flux in direction x and $\frac{\partial T}{\partial x}$ is temperature field in direction x.

The differential equation of heat conduction can be obtained according to Fourier's law and energy conservation law, and it is hypothesized that there is internal heat source in the object. Based on the above hypothesis, microelement

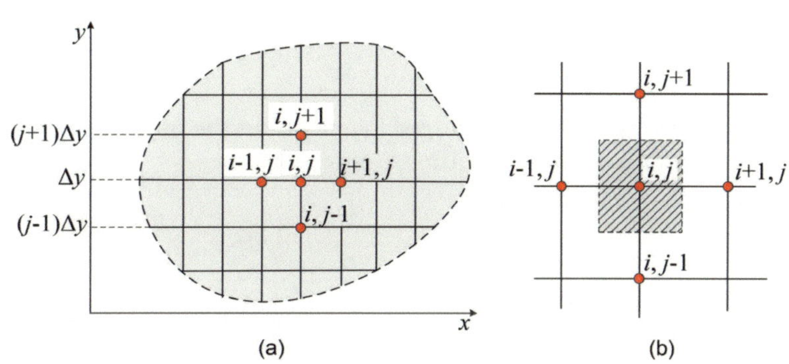

Figure 2.
Schematic diagram of gridlines and nodes of finite difference method. (a) Grid element. (b) Grid node.

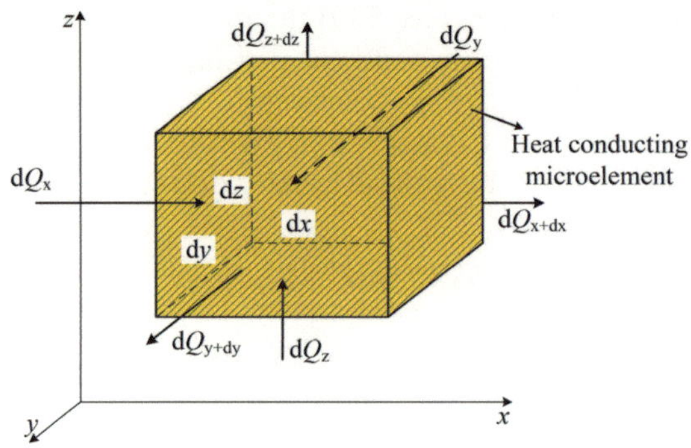

Figure 3.
Schematic diagram of thermal conductive microelement.

dV = dx is divided from the object under heat conduction. As shown in **Figure 3**, the three edges of this microelement are parallel to axes x, y, and z, respectively. Thermal equilibrium analysis of the microelement is conducted. It can be known from the energy conservation law that the net heat quantity transferred in and out of the microelement within time dt should be equal to increment of internal energy in the microelement, namely net heat quantity transferred in and out of the microelement (I) and increment of internal energy in the microelement (II).

Net heat quantity transferred in and out of the microelement and increment of internal energy in the microelement will be, respectively, calculated as follows.

Energy equilibrium analysis is implemented for the microelement in **Figure 3**. The net quantity transferred in and out of the microelement can be obtained by adding the net heat quantities transferred in and out of the microelement from directions x, y, and z, respectively. Along the direction of axis x, the heat quantity transferred in the microelement within time dt through plane x is:

$$\mathrm{d}Q_{\mathrm{x}} = q_{\mathrm{x}}\mathrm{d}y\mathrm{d}z\mathrm{d}t \tag{4}$$

The heat quantity transferred out through plane x + dx:

$$\mathrm{d}Q_{\mathrm{x+dx}} = q_{\mathrm{x+dx}}\mathrm{d}y\mathrm{d}z\mathrm{d}t \tag{5}$$

where $q_{\mathrm{x+dx}} = q_{\mathrm{x}} + \frac{\partial q_{\mathrm{x}}}{\partial x}dx$.

Hence, the net heat quantity transferred in and out of the microelement along the direction of axis x within time dt is:

$$\mathrm{d}Q_{\mathrm{x}} - \mathrm{d}Q_{\mathrm{x+dx}} = -\frac{\partial q_{\mathrm{x}}}{\partial x}\mathrm{d}x\mathrm{d}y\mathrm{d}z\mathrm{d}t \tag{6}$$

Similarly, the net quantities transferred in and out of the microelement along the directions of axes y and z within this time are, respectively:

$$\mathrm{d}Q_{\mathrm{y}} - \mathrm{d}Q_{\mathrm{y+dy}} = -\frac{\partial q_{\mathrm{y}}}{\partial y}\mathrm{d}x\mathrm{d}y\mathrm{d}z\mathrm{d}t \tag{7}$$

$$\mathrm{d}Q_{\mathrm{z}} - \mathrm{d}Q_{\mathrm{z+dz}} = -\frac{\partial q_{\mathrm{z}}}{\partial z}\mathrm{d}x\mathrm{d}y\mathrm{d}z\mathrm{d}t \tag{8}$$

The following is obtained by adding net quantities transferred in and out of the microelement in directions x, y, and z:

$$\mathrm{I} = -\left(\frac{\partial q_{\mathrm{x}}}{\partial x} + \frac{\partial q_{\mathrm{y}}}{\partial y} + \frac{\partial q_{\mathrm{z}}}{\partial z}\right)\mathrm{d}x\mathrm{d}y\mathrm{d}z\mathrm{d}t \tag{9}$$

Within time dt, the increment of internal energy in the microelement is:

$$\mathrm{II} = \rho c\frac{\partial T}{\partial t}\mathrm{d}x\mathrm{d}y\mathrm{d}z\mathrm{d}t \tag{10}$$

where ρ is density of point heat source heat conducting medium and c is the specific heat.

The three-dimensional heat conduction model can be obtained by Eqs. (9) and (10):

$$\frac{\partial T}{\partial t} = \alpha_{\mathrm{t}}\left(\frac{\partial^2 T}{\partial x^2} + \frac{\partial^2 T}{\partial y^2} + \frac{\partial^2 T}{\partial z^2}\right) \tag{11}$$

where α_{t} is thermal diffusion coefficient of point heat source heat conducting medium.

2.1.2.3 *Differential equation converted into difference equation*

This specimen is assumed as a rectangular plane, and it is discretely decomposed into a planar grid structure. Isometric spatial step length $\Delta x = \Delta z = \Delta l$ is taken; two groups of equally spaced parallel lines are drawn to subdivide the rectangular specimen and the equation of parallel line:

$$\begin{cases} x = x_i = i\Delta l, i = 0, 1, \ldots, M, M\Delta l = l_w \\ z = z_j = j\Delta l, j = 0, 1, \ldots, N, N\Delta l = b_w \end{cases} \tag{12}$$

where x_{i} and z_{j} are coordinate value of transverse line i in direction x and coordinate value of vertical line j in direction z, respectively, and both lines constitute the difference grid; l_{w} and b_{w} are specimen length and height, respectively; and M and N are natural numbers.

The grid area of difference calculation is obtained through subdivision as shown in **Figure 3**.

A set of difference equations is built based on second-order difference quotient, namely:

$$\begin{cases} \dfrac{\partial^2 T}{\partial x^2}(i,j) = \dfrac{T(i+1,j) + T(i-1,j) - 2T(i,j)}{\Delta l^2} + \mathrm{O}(\Delta l^2) \\ \dfrac{\partial^2 T}{\partial z^2}(i,j) = \dfrac{T(i,j+1) + T(i,j-1) - 2T(i,j)}{\Delta l^2} + \mathrm{O}(\Delta l^2) \\ \dfrac{\partial T}{\partial t}(i,j) = \dfrac{T_{t+\Delta t}(i,j) - T_{\mathrm{t}}(i,j)}{\Delta t} + \mathrm{O}(\Delta t) \end{cases} \tag{13}$$

The difference equation of each node in the grid can be obtained:

$$T_{\mathrm{t+\Delta t}}(i,j) = \frac{\Delta t\{k_{\mathrm{x}} \cdot [T(i,j+1) + T(i,j-1)] + k_{\mathrm{z}} \cdot [T(i+1,j) + T(i-1,j)]\}}{\rho_{\mathrm{w}} c_{\mathrm{w}} \Delta l^2} + \left[1 - \frac{2\Delta t(k_{\mathrm{x}} + k_{\mathrm{z}})}{\rho_{\mathrm{w}} c_{\mathrm{w}} \Delta l^2}\right] T_{\mathrm{t}}(i,j) \tag{14}$$

2.2 Boundary condition

Heat transfer means transferring heat quantity from a system to another system with three main forms: convection, heat conduction, and radiation [25, 26]. The grinding tool/specimen contact interface is cooled with grinding fluid in the grinding process, so the grinding fluid flowing at this interface will conduct heat transfer on the specimen surface, where the main heat transfer form is convective heat transfer. As the grinding fluid contacts the specimen surface, partial heat quantity on the specimen surface is brought away by cooling fluid via convective heat transfer, and the residual heat stays on the specimen matrix and is transferred inside it. Therefore, the cooling effect of the machining zone can be strengthened by cooling fluid or by enhancing its heat transfer performance [27]. The temperature of external medium and convective heat transfer coefficient on boundary are given as shown in **Figure 4**. The heat quantity at grinding interface is generated due to grinding action of grinding tool and brought away by external cooling medium. In the meantime, both cooling medium and adjacent node (i-1, 1) on the specimen surface contact this node (i, 1), so the heat quantity at this node will be transferred to cooling medium in convective way and also to adjacent node, and in the end, a stable temperature is reached at grinding tool/specimen interface [28].

The thermal equilibrium analysis of heat conduction in temperature field is to solve differential equation. The complete description of heat conduction process includes monodromy conditions and differential equation of heat conduction, where the former includes time condition, physical condition, geometrical condition, and boundary condition [29]. The boundary condition specifies characteristics of heat transfer process on object boundary, namely reflecting conditions for interaction between heat transfer process and surroundings, so initial condition (temperature at t = 0) and boundary conditions (set temperature in boundary region or input and output heat fluxes) must be set in the temperature field. Based on the heat transfer theory, thermodynamic boundary conditions are generally divided into the three following types [30].

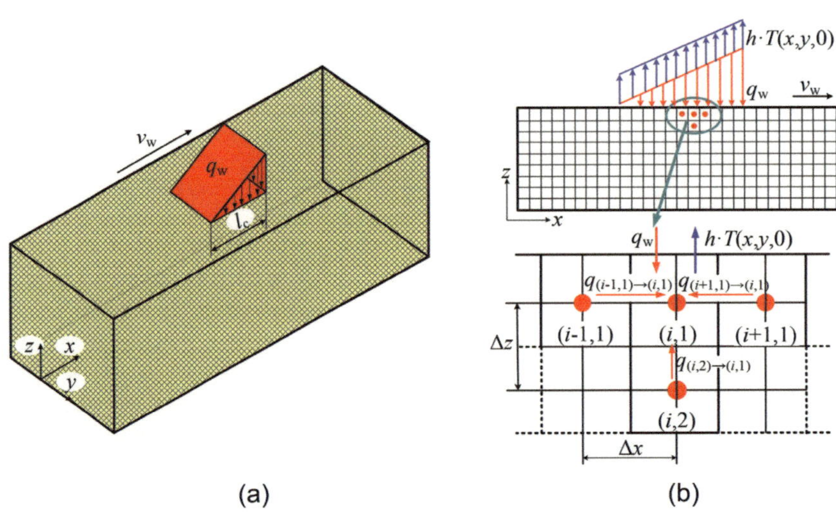

Figure 4.
Schematic diagram of heat conduction model and convective heat transfer at grinding interface. (a) Heat conduction model. (b) Convective heat transfer at grinding interface.

2.2.1 The first kind of boundary conditions

Boundary condition of this type is forced convective boundary condition, namely temperature value at boundary of the given object at any time, and it is also called Dirichlet condition:

$$T|_{s_f} = T_w \tag{15}$$

where T_w is set temperature at boundary surface s_f. If boundary temperature is kept unchanged, T_w is a fixed value. If boundary temperature changes with time, T is a time-related functional expression.

2.2.2 The second kind of boundary conditions

Boundary condition of this type refers to heat flux on object boundary surface in normal direction at any time, and it is also called Neumann condition. The relationship between temperature gradient and heat flux is obtained through Fourier's law, and this is equivalent to temperature change rate on boundary s_f in normal direction at any time:

$$\left.\frac{\partial T}{\partial n}\right|_{s_f} = \frac{Q_w}{k} \tag{16}$$

where Q_w is heat quantity passing through boundary surface s_f. $Q_w = 0$ when the boundary is under thermal insulation. Q_w is a fixed value when heat conduction at the boundary is constant, and it is a time-related function when the heat conduction changes with time.

Differential equation at the second type of boundary:

$$\frac{\partial}{\partial x}\left(k_x \cdot \frac{\partial T}{\partial x}\right) + \frac{\partial}{\partial y}\left(k_y \cdot \frac{\partial T}{\partial y}\right) + \frac{\partial}{\partial z}\left(k_z \cdot \frac{\partial T}{\partial z}\right) = Q_w \tag{17}$$

2.2.3 The third kind of boundary conditions

Boundary condition of this type denotes convective heat transfer between boundary surface and surrounding medium, and it is also called Robin condition. It can be known from Newton's law of cooling that convective heat transfer occurs between boundary layer of the specimen and cooling heat transfer medium:

$$h(T|_{z=0} - T_0) - k\left.\frac{\partial T}{\partial z}\right|_{z=0} = q_w \tag{18}$$

where q_w is convective heat flux at the boundary surface s_f between cooling heat transfer medium and specimen and h is convective heat transfer coefficient at the boundary between cooling heat transfer medium and specimen.

Differential equation at the third type of boundary:

$$\frac{\partial}{\partial x}\left(k_x \cdot \frac{\partial T}{\partial x}\right) + \frac{\partial}{\partial y}\left(k_y \cdot \frac{\partial T}{\partial y}\right) + \frac{\partial}{\partial z}\left(k_z \cdot \frac{\partial T}{\partial z}\right) = h\left(T|_{s_f} - T_f\right) \tag{19}$$

2.3 Heat distribution coefficients

An important problem exists in the establishment process of temperature field model in the grinding zone. It is necessary to determine the proportion of grinding heat quantity transferred into the specimen, namely heat distribution coefficient R_w. Under non-dry grinding condition as shown in **Figure 4**, the total heat quantity generated in the grinding zone is [31]:

$$q_{\text{total}} = \frac{F_t v_s}{b_g l_c} = q_w + q_g + q_c + q_f \quad (20)$$

where b_g is width of grinding tool; q_w is heat quantity staying on the specimen surface; q_g is heat quantity transferred into abrasive particle of grinding tool; q_c is heat quantity carried away by grinding chips; and q_f is heat quantity transferred out by cooling medium.

The heat distribution coefficient model proposed by Rowe in consideration of convective heat transfer in the grinding zone is generally applied. As indicated by the model, under high-efficiency deep grinding, total energy in the grinding zone is transferred into grinding tool, grinding chips, grinding fluid, and theoretical mathematical model of specimen material, and furthermore, its feasibility has been proved by a large quantity of experimental results.

Heat flux transferred into specimen, grinding tool, grinding fluid, and grinding chips is correlated with parameters like maximum contact temperature $T_{\max}$, boiling point T_b of grinding fluid, and melting point T_m of specimen ($T_{\max} \le T_b$) as shown in the following formula [32]:

$$\begin{cases} q_w = h_w \cdot T_{max} \\ q_g = h_g \cdot T_{max} \\ q_f = h_f \cdot T_{max} \\ q_c = h_d \cdot T_m \end{cases} \quad (21)$$

where h_w, h_g, h_f, and h_d are heat transfer coefficients of specimen material, grinding tool, grinding fluid, and grinding chips, respectively.

The proportion of heat quantity flowing into the specimen can be obtained as follows:

$$R_w = h_w \cdot T_{\max}/q_{\text{total}} \quad (22)$$

3. Experimental study on micro grinding of biological bone with nanoparticle jet mist cooling

3.1 Nanoparticle jet mist cooling bone micro grinding experimental platform

Figure 5 shows the established experimental platform of NJMC bio-bone micro grinding, including feed system, fixed system, cooling system, and measuring system. Both axes x and y of the 3D displacement device are Shinano stepper motors, and axis z is Panasonic servo braking motor. A 120# diamond grinding tool is used, where the diameter of grinding head is 1 mm. The mist cooling supply device is mist

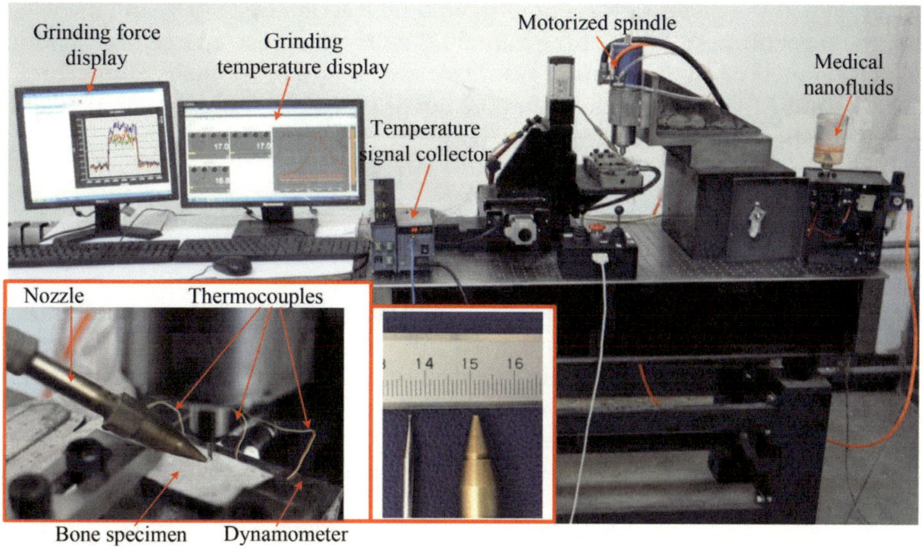

Figure 5.
Nanoparticle jet mist cooling bone micro grinding experimental platform.

cooling supply system produced by Shanghai KINS Energy-Saving Technology Co., Ltd. The jet parameters are uniformly set in the experiment as shown in **Table 1**.

The clamping mode of thermocouple: a blind hole drilled on the back face of the bone specimen has a certain distance from the bone surface (grinding depth), the thermocouple wire is placed into the blind hole, the bone specimen is fixed on the dynamometer, and the force and temperature in the micro bone grinding process are simultaneously measured. Thermocouple type is K (TT-K-30), and measuring frequencies of both dynamometer and thermocouple are 100 Hz.

In the biomedical field, HA, SiO_2, and Al_2O_3 nanoparticles feature nontoxicity and good biological compatibility and are commonly used drug carriers in nanodrug release system [33, 34]; NS is a common clinically used cooling medium as osmotic pressure is basically equal to osmotic pressure of human plasma. Therefore, HA, SiO_2, and Al_2O_3 nanoparticles with a diameter of 50 nm were used as solid nanoscale additives and NS as a nanofluid base to prepare HA, SiO_2, and Al_2O_3 nanofluids. Polyethylene glycol 400 has been extensively applied to lubrication in colonoscopy and gastroscopy by virtue of superior lubricating property and nontoxicity, and its safety in human body has been clinically certified. In the meantime, PEG400 also has good dispersity. Therefore, PEG400 was used as dispersing agent in this

Jet parameters	Value
Nozzle diameter	d_0 = 1 mm
Compressed gas pressure	Δp = 0.5 MPa
Feed flow rate	Q_f = 50 mL/h
Spray inclination angle	β = 15°
Nozzle height	H = 0.6 mm
Spray cone angle	α = 27°

Table 1.
Table of mist cooling jet parameters.

chapter. When the dispersing agent with volume fraction of 2 vol.% and 2 vol.% is used, the suspension stability of the nanofluid will be the best. Hence, the nanofluid preparation method in this chapter was "two-step method," namely adding 2 mL of HA, SiO_2, and Al_2O_3 nanoparticles and 0.2 mL of PE in 100 mL of NS supplemented with ultrasonic vibration for 15 min.

3.2 Temperature field of bio-bone micro grinding with different cooling methods

The dynamic temperature field model of NJMC bone grinding was verified. Different cooling modes were adopted: dry grinding, mist, and NJMC. As dynamic heat flux was loaded to theoretically calculate grinding temperature field, temperatures at different measuring points on the surface of the bone material were measured. As shown in **Figure 6**, three groups of thermocouples were used to simultaneously measure the temperatures at three measuring points—T1, T2, and T3—with spacing of 5 mm.

Figure 7(a) shows typical grinding force signals measured under dry grinding condition, where F_y is the force used to generate heat quantity.

The temperature field of bone micro grinding is solved under initial condition (room temperature) of T_0 = 20°C. **Figure 8(a)** shows temperature curves measured at different measuring points under dry grinding condition and theoretical temperature curves. It can be known that in the bone grinding temperature curves, the peak values at three measuring points are 36.1°C, 38.2°C, and 36°C, respectively, and measured peak values at the three measuring points are 36.7°C, 38.5°C, and 36.6°C, respectively, namely the temperature on bone surface is ever-changing.

Figure 8(b) and **(c)** show peak values measured through the experiment at three measuring points under mist and NJMC conditions and those obtained through calculation. It can be known that in comparison with average temperature (36.8°C) on bone surface under dry grinding condition, the temperatures on the bone surface under mist and NJMC decline by 13.6% and 33.9%, respectively, thus proving the superior cooling effect of NJMC mode. The temperature error is 6.7%, and theoretical analysis basically accords with experimental result, which verifies the correctness of the dynamic temperature field in NJMC bio-bone micro grinding [35].

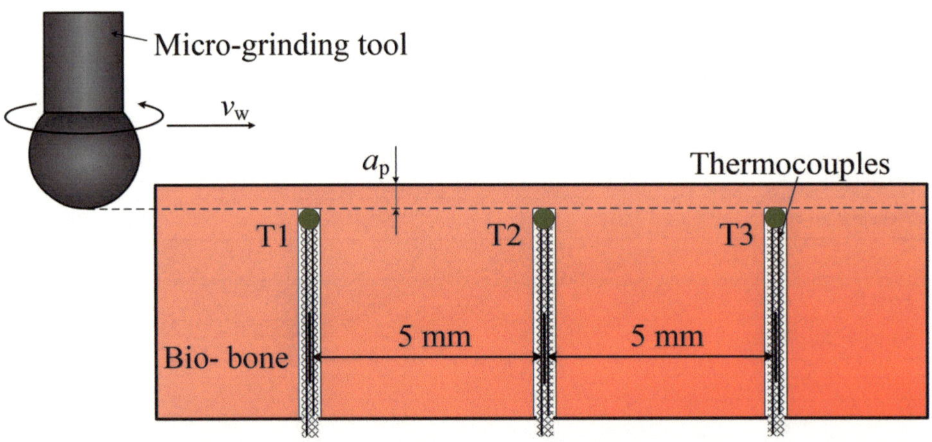

Figure 6.
Schematic diagram of thermocouple clamping at different measuring points on bone surface.

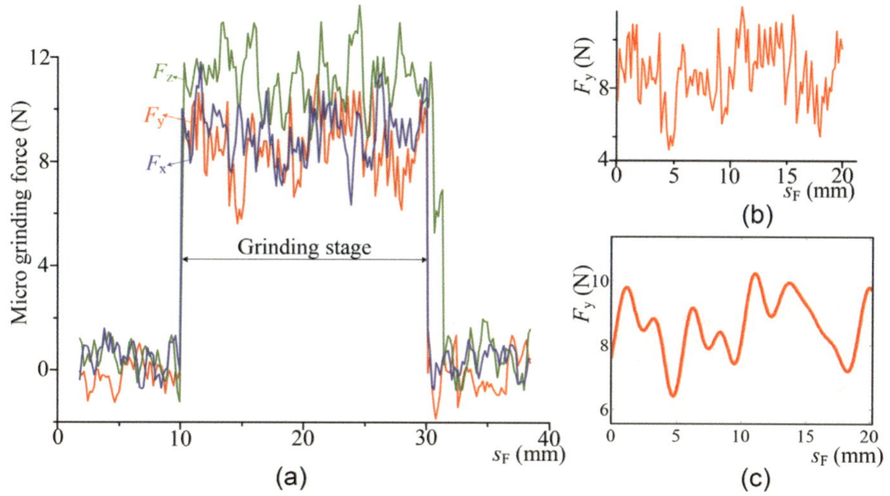

Figure 7.
Typical force signals in micro grinding of biological bone materials. (a) Typical force signal of bio-bone micro grinding. (b) Grinding force in y direction. (c) Fitting curve of grinding force in y direction.

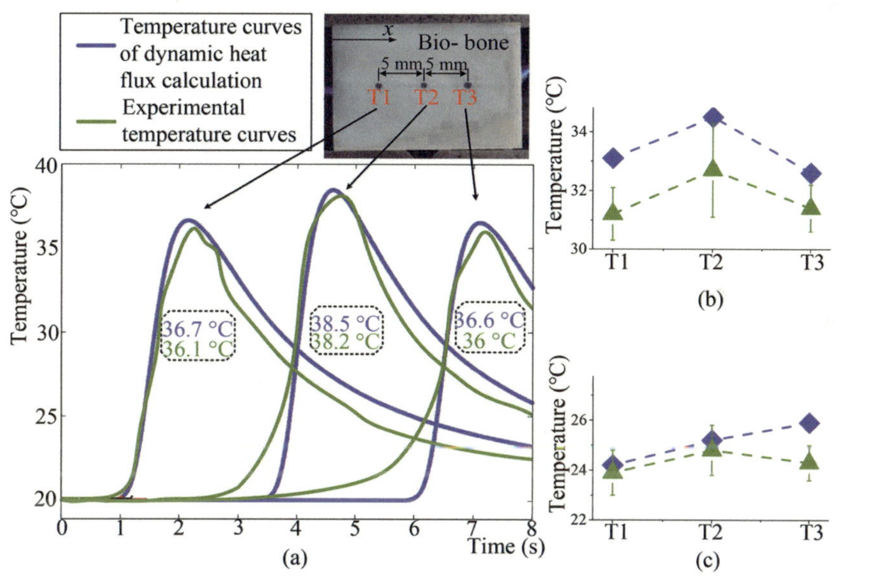

Figure 8.
Temperature curves and values of different measuring points on bone surface under different cooling conditions. (a) Dry grinding temperature. (b) Mist cooling grinding temperature. (c) NJMC grinding temperature.

In order to explore into the dynamic characteristics of temperature field in bio-bone micro grinding, temperature fields in cut-in zone, steady-state zone, and cut-out zone are, respectively, analyzed. **Figures 9** and **10** show temperature fields and temperature curves in bone grinding under NJMC conditions. The detailed analysis process is as follows:

1. Cut-in zone: As shown in **Figures 9(a)** and **10**, as the grinding starts, the grinding tool starts contacting the material and gradually cuts into the material, undeformed cutting thickness is gradually increasing [36, 37], and the heat quantity generated at grinding interface starts migrating into the

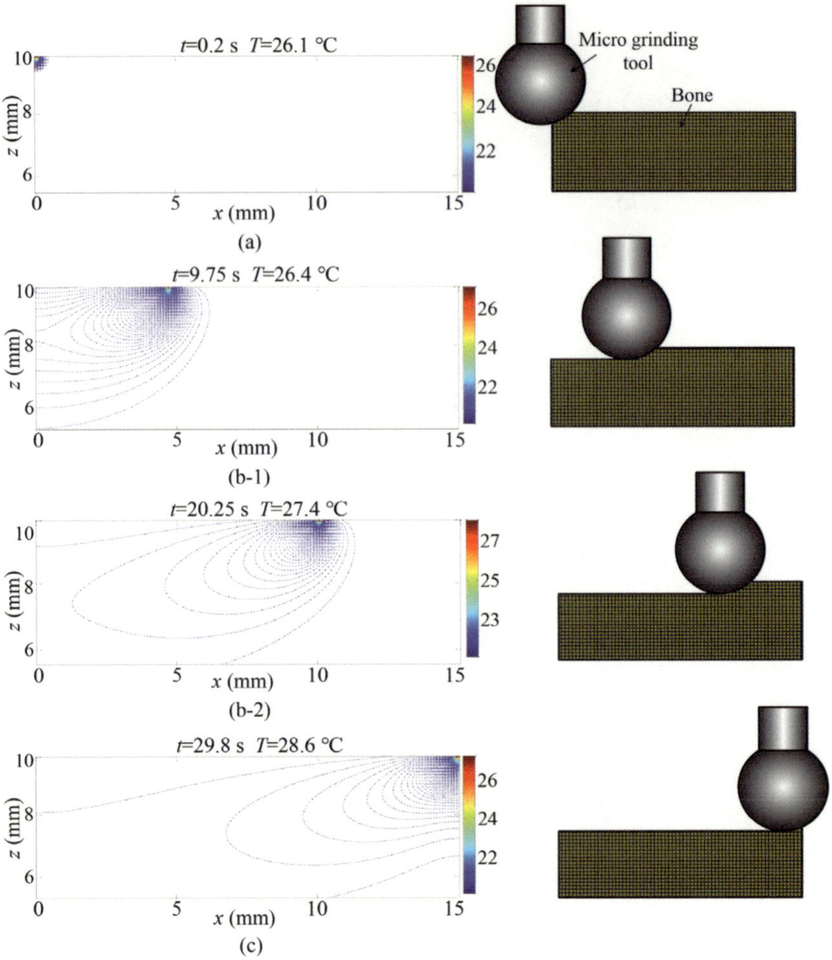

Figure 9.
Temperature field in cut-in, steady-state, and cut-out zones of bone micro grinding. (a) Cut-in temperature field. (b-1) Steady-state temperature field. (b-2) Steady-state temperature field. (c) Cut-out temperature field.

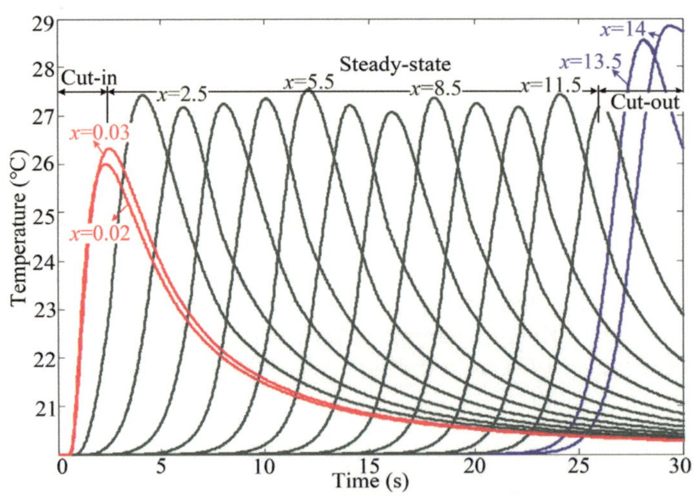

Figure 10.
Temperature curves of cut-in, steady-state, and cut-out zones in bone micro grinding.

specimen surface. The temperature is low in the grinding zone, because the volume of material participating in the grinding in the cut-in zone is small with a small energy consumption.

2. Steady-state zone: In **Figures 9(b), (c)** and **10**, undeformed cutting thickness is kept at average value, and surface temperature no longer continues to rise, namely the formed temperature field reaches a steady state. As for grinding temperature fields calculated by researchers previously, after constant heat flux is loaded, the temperature value under which a steady state is reached will not be changed. However, in the temperature field of bone micro grinding with loaded dynamic heat flux as shown in the figure, the temperature value is ever-changing after the steady state is reached [38, 39].

3. Cut-out zone: As shown in **Figures 9(d)** and **10**, according to the heat transfer theory, undeformed thickness is gradually reduced when the grinding tool is in cut-out zone, and the volume of specimen material participating in the grinding action is continuously reduced. If the heat quantity generated at grinding interface remains unchanged, the volume of the material to which heat is migrated in the cut-out zone is continuously reduced. As heat conductivity coefficient of air is extremely low, more heat is aggregated in the grinding zone, and then grinding temperature rises to a great extent.

3.3 Effect of nanoparticle size on bone micro grinding temperature

To probe into the influence laws of nanoparticle size on bone micro grinding temperature, Al_2O_3 nanoparticles with particle sizes of 30, 50, 70, and 90 nm and NS were used to prepare nanofluids for bone grinding experiment and measure grinding force and temperature at measuring point T2 on the bone surface.

Figure 11(a) shows theoretical temperature curves under different cooling conditions, where abscissa axis denotes dimensionless distance ($x/l = 2v_w t/l_c$, $2l = l_c$), namely the location of this point in the grinding arc zone [40, 41]. It can be known from the figure that at the entry end in the contact zone, grinding temperature abruptly increases, peak temperature deviates from central position of heat source on the curve and deflects toward exit end of the contact zone, and the temperature at the exit end in the contact zone declines slowly. **Figure 11(b)** shows bone grinding temperatures measured using nanoparticles with different particle sizes, and it can be known that the grinding temperature increases with nanoparticle size.

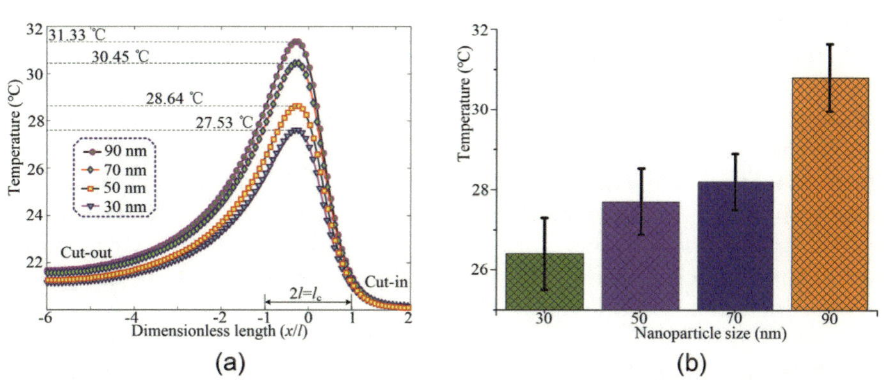

Figure 11.
Micro grinding temperature with different nanoparticle size. (a) Theoretical calculated temperature value (b) theoretical calculated temperature value.

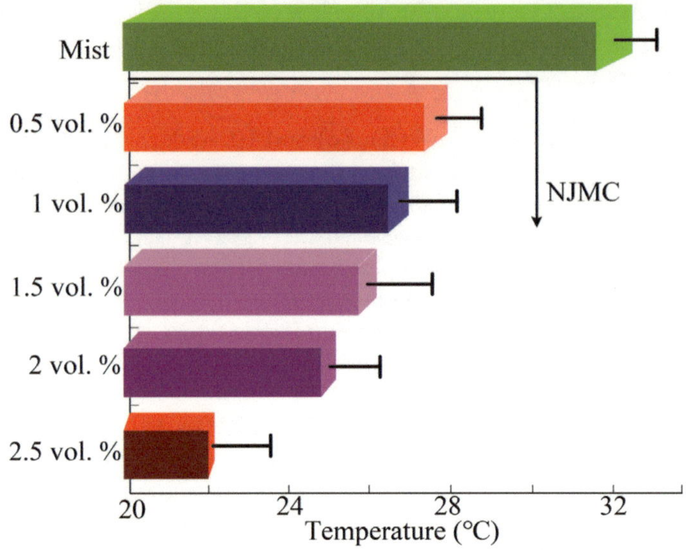

Figure 12.
Temperature values measured with different concentration of nanoparticles.

3.4 Effect of nanoparticle concentration on bone micro grinding temperature

SiO_2 nanofluid with volume fraction of 0.5, 1, 1.5, 2, and 2.5 vol.% was prepared in the experiment to investigate the influence laws of nanoparticle concentration on bone micro grinding temperature. Mist cooling was taken for comparative experiment and to measure grinding force and temperature at measuring point T2.

As shown in **Figure 12**, the temperature measured under mist cooling is 32.7°C. Mist cooling is taken for control, and the surface temperature obtained using nanofluid with volume fraction of 0.5, 1, 1.5, 2, and 2.5 vol.% declines by 14.1%, 17.1%, 19.6%, 22.9%, and 33.3%, respectively, namely the surface temperature in micro grinding decreases as the nanoparticle volume fraction increases.

4. Conclusions

As irreversible thermal injury and poor visibility of operative region exist in the current clinical bone micro grinding operation, a NJMC bio-bone micro grinding process has been proposed to investigate the convective heat transfer mechanism of nanofluid in the grinding zone and reveal the heat distribution mechanism. On this basis, a dynamic temperature field model in NJMC bio-bone micro grinding process has been established. An experimental platform of NJMC bio-bone micro grinding process has been set up to realize an experimental verification of dynamic temperature fields in this grinding process, followed by an experimental study of influence of laws of nanoparticle size and concentration on bone grinding temperature. The following conclusions are drawn:

1. Heat distribution coefficient of specimen material, that of abrasive particle, that of grinding chips, and that of cooling medium, characterizes the abilities of heat transfer media to contend for heat quantity in the unit area of the grinding zone within unit time, so the heat distribution coefficient of the

specimen material can be expressed as the function of heat transfer coefficient of each heat transfer medium in the micro grinding zone.

2. Compared with average temperature (36.8°C) on bone surface under dry grinding condition, bone surface temperatures under mist cooling and NJMC conditions decrease by 13.6% and 33.9%, respectively, thus verifying the superior cooling effect of NJMC. The temperature error is 6.7%, theoretical analysis basically accords with experimental result, and thus the correctness of temperature field in NJMC bio-bone micro grinding is proved.

3. Al_2O_3 nanoparticles with different particle sizes and NS were employed to prepare nanofluids for bone grinding experiment. According to the experimental results, the grinding temperature rose abruptly at entry end of the contact zone, the peak temperature value deviated from the central position of heat source on the curve and deflected toward the exit end of the contact zone, the temperature at the exit end of the contact zone declined slowly, and the grinding temperature increased with the nanoparticle size.

4. The bone grinding experiment has been carried out with nanofluids prepared using SiO_2 nanoparticles with different volume fractions and NS. In comparison with mist cooling, the surface temperatures obtained through nanofluids with nanoparticle volume fraction of 0.5, 1, 1.5, 2, and 2.5 vol.% declined by 14.1%, 17.1%, 19.6%, 22.9%, and 33.3%, respectively, namely the surface temperature declined as the nanoparticle volume fraction increased in grinding.

Acknowledgements

This research was financially supported by the following organizations: the National Natural Science Foundation of China (51975305 and 51905289), the Major Research Project of Shandong Province (2019GSF108236, 2019GGX104040, and 2018GGX103044), the Shandong Provincial Natural Science Foundation of China (ZR2019PEE008), Major Science and Technology Innovation Engineering Projects of Shandong Province (2019JZZY020111), and Applied Basic Research Youth Project of Qingdao Science and Technology Plan (19-6-2-63-cg).

Conflict of interests

The authors confirm that no conflict of interest exists in this chapter.

Appendices and nomenclature

NJMC	nanoparticle jet mist cooling
T_w	set temperature at boundary surface
NS	normal saline
Q_w	heat quantity passing through boundary surface
3D	three-dimensional
q_w	convective heat flux at the boundary surface
T	grinding temperature
h	convective heat transfer coefficient at the boundary between cooling heat transfer medium and specimen

x, y, z	spatial coordinates
b_g	width of grinding tool
t	time
q_w	heat quantity staying on the specimen surface
l_c	contact arc length
q_g	heat quantity transferred into abrasive particle of grinding tool
Q	the heat quantity passing through infinitesimal isothermal surface
q_c	heat quantity carried away by grinding chips
q_x	heat flux in direction x
q_f	heat quantity transferred out by cooling medium
ρ	density of point heat source heat conducting medium
T_{max}	maximum contact temperature
c	specific heat of point heat source heat conducting medium
T_b	boiling point of grinding fluid
α_t	thermal diffusion coefficient of point heat source heat conducting medium
T_m	melting point of specimen
l	isometric spatial step length
h_w	heat transfer coefficients of specimen material
l_w	specimen length
h_g	heat transfer coefficients of grinding tool
b_w	specimen height
h_f	heat transfer coefficients of grinding fluid
M, N	natural numbers
h_d	heat transfer coefficients of grinding chips
s_f	convection heat transfer boundary surface
R_w	the proportion of heat quantity flowing into the specimen

Author details

Min Yang[1], Changhe Li[1*], Liang Luo[2], Lan Dong[3], Dongzhou Jia[4], Runze Li[5], Mingzheng Liu[1], Xin Cui[1], Yali Hou[1], Yanbin Zhang[1], Teng Gao[1], Xiaoming Wang[1] and Yunze Long[6*]

1 School of Mechanical and Automotive Engineering, Qingdao University of Technology, Qingdao, China

2 Ningbo Sanhan Alloy Material Co., Ltd, Ningbo, China

3 School of Mechanical and Electrical Engineering, Qingdao Binhai University, Qingdao, China

4 School of Mechanical Engineering, Inner Mongolia University for Nationalities, Tongliao, China

5 Department of Biomedical Engineering, University of Southern California, Los Angeles, United States

6 School of Physical Sciences, Qingdao University, Qingdao, China

*Address all correspondence to: sy_lichanghe@163.com and yunze.long@163.com

References

[1] Lee PH, Lee SW, Lim SH, et al. A study on thermal characteristics of micro-scale grinding process using nanofluid minimum quantity lubrication (MQL). International Journal of Precision Engineering and Manufacturing. 2015;**16**(9):1899-1909. DOI: 10.1007/s12541-015-0247-2

[2] Feldmann A, Ganser P, Nolte L, et al. Orthogonal cutting of cortical bone: Temperature elevation and fracture toughness. International Journal of Machine Tools and Manufacture. 2017; **118-119**:1-11. DOI: 10.1016/j.ijmachtools.2017.03.009

[3] Zhang L, Tai BL, Wang AC, et al. Mist cooling in neurosurgical bone grinding. CIRP Annals - Manufacturing Technology. 2013;**62**(1):367-370. DOI: 10.1016/j.cirp.2013.03.125

[4] Jia DZ, Li CH, Zhang YB, et al. Experimental evaluation of surface topographies of NMQL grinding ZrO_2 ceramics combining multiangle ultrasonic vibration. The International Journal of Advanced Manufacturing Technology. 2019;**100**(1–4):457-473. DOI: 10.1007/s00170-018-2718-y

[5] Yang M, Li CH, Zhang YB, et al. Effect of friction coefficient on chip thickness models in ductile-regime grinding of zirconia ceramics. The International Journal of Advanced Manufacturing Technology. 2019; **102**(5):2617-2632. DOI: 10.1007/s00170-019-03367-0

[6] Yang M, Li CH, Zhang YB, et al. Predictive model for minimum chip thickness and size effect in single diamond grain grinding of zirconia ceramics under different lubricating conditions. Ceramics International. 2019;**45**(12):14908-14920. DOI: 10.1016/j.ceramint.2019.04.226

[7] Yang M, Li CH, Zhang YB, et al. Maximum undeformed equivalent chip thickness for ductile-brittle transition of zirconia ceramics under different lubrication conditions. International Journal of Machine Tools and Manufacture. 2017;**122**:55-65. DOI: 10.1016/j.ijmachtools.2017.06.003

[8] Zhang YB, Li CH, Yang M, et al. Experimental evaluation of cooling performance by friction coefficient and specific friction energy in nanofluid minimum quantity lubrication grinding with different types of vegetable oil. Journal of Cleaner Production. 2016;**139**: 685-705. DOI: 10.1016/j.jclepro.2016.08.073

[9] Wu WT, Li CH, Yang M, et al. Specific energy and g ratio of grinding cemented carbide under different cooling and lubrication conditions. The International Journal of Advanced Manufacturing Technology. 2019;**105** (1-4):67-82. DOI: 10.1007/s00170-019-04156-5

[10] Jia DZ, Li CH, Zhang YB, et al. Experimental research on the influence of the jet parameters of minimum quantity lubrication on the lubricating property of Ni-based alloy grinding. International Journal of Advanced Manufacturing Technology. 2016;**82**: 617-630. DOI: 10.1007/s00170-015-7381-y

[11] Zhang YB, Li CH, Ji HJ, et al. Analysis of grinding mechanics and improved predictive force model based on material-removal and plastic-stacking mechanisms. International Journal of Machine Tools and Manufacture. 2017;**122**: 81-97. DOI: 10.1016/j.ijmachtools.2017.06.002

[12] Cui X, Li CH, Zhang YB, et al. Tribological properties under the grinding wheel and workpiece interface by using graphene nanofluid lubricant. The International Journal of Advanced

Manufacturing Technology. 2019;**104**(9–12):3943-3958

[13] Zhang LH, Tai BL, Wang GJ, et al. Thermal model to investigate the temperature in bone grinding for skull base neurosurgery. Medical Engineering & Physics. 2013;**35**(10):1391-1398. DOI: 10.1016/j.medengphy.2013.03.023

[14] Zhu Z, Hu ZW, Zhang ZB, et al. Study on grinding force of bovine cortical bone by different grinding wheels. Diamond & Abrasives Engineering. 2014;**34**(5):13-16

[15] Sasaki M, Morris S, Goto T, et al. Spray-irrigation system attached to high-speed drills for simultaneous prevention of local heating and preservation of a clear operative field in spinal surgery. Neurologia Medico-Chirurgica. 2010;**50**(10):900-904. DOI: 10.2176/nmc.50.900

[16] Enomoto T, Shigeta H, Sugihara T, et al. A new surgical grinding wheel for suppressing grinding heat generation in bone resection. CIRP Annals - Manufacturing Technology. 2014;**63**(1): 305-308. DOI: 10.1016/j.cirp.2014.03.026

[17] Zhang XM, Ren ZP, Mei FM. Heat Transfer. Beijing: China Construction Industry Press; 1995. pp. 75-96

[18] Jaeger JC. Moving sources of heat and the temperature of sliding contacts. Proceedings of the Royal Society of New South Wales. 1942;**76**:203-224

[19] Li HN, Xie KG, Wu B, et al. Generation of textured diamond abrasive tools by continuous-wave CO_2 laser: Laser parameter effects and optimisation. Journal of Materials Processing Technology. 2020;**275**: 116279. DOI: 10.1016/j.jmatprotec.2019.116279

[20] Li HN, Yang Y, Zhao YJ, et al. On the periodicity of fixed-abrasive planetary lapping based on a generic model. Journal of Manufacturing Processes. 2019;**44**:271-287. DOI: 10.1016/j.jmapro.2019.05.036

[21] Yang M, Li CH, Zhang YB, et al. Experimental research on microscale grinding temperature under different nanoparticle jet minimum quantity cooling. Materials and Manufacturing Processes. 2016;**32**:589-597. DOI: 10.1080/10426914.2016.1176198

[22] Gao T, Zhang XP, Li CH, et al. Surface morphology evaluation of multi-angle 2D ultrasonic vibration integrated with nanofluid minimum quantity lubrication grinding. Journal of Manufacturing Processes. 2020;**51**:44-61. DOI: 10.1016/j.jmapro.2020.01.024

[23] Zhang DK, Li CH, Zhang YB, et al. Experimental research on the energy ratio coefficient and specific grinding energy in nanoparticle jet MQL grinding. International Journal of Advanced Manufacturing Technology. 2015;**78**:1275-1288. DOI: 10.1007/s00170-014-6722-6

[24] Shen B. Minimum Quantity Lubrication Grinding Using Nanofluids. Michigan: The University of Michigan; 2008

[25] Malkin S, Guo C. Thermal analysis of grinding. CIRP Annals - Manufacturing Technology. 2007;**56**(2):760-782

[26] Kuwahara F, Shirota M, Nakayama A. A numerical study of interfacial convective heat transfer coefficient in two-energy equation model for convection in porous media. International Journal of Heat and Mass Transfer. 2001;**44**(6):1153-1159. DOI: 10.1016/s0017-9310(00)00166-6

[27] Duursma G, Sefiane K, Kennedy A. Experimental studies of nanofluid droplets in spray cooling. Heat Transfer Engineering. 2009;**30**(13):1108-1120. DOI: 10.1080/01457630902922467

[28] Zhang DK, Li CH, Jia DZ, et al. Specific grinding energy and surface roughness of nanoparticle jet minimum quantity lubrication in grinding. Chinese Journal of Aeronautics. 2015;**28**(2): 570-581. DOI: 10.1016/j.cja.2014.12.035

[29] Li HN, Axinte D. On a stochastically grain-discretised model for 2D/3D temperature mapping prediction in grinding. International Journal of Machine Tools and Manufacture. 2017; **116**:60-76. DOI: 10.1016/j.ijmachtools.2017.01.004

[30] Yang SM, Tao WS. Heat Transfer. Higher Education Press: Beijing; 1998. pp. 33-296p

[31] Outwater JQ, Shaw MC. Surface temperature in grinding. ASME. 1952; **12**(1):73-78

[32] Rowe WB, Black SCE, Mills B, et al. Analysis of grinding temperatures by energy partitioning. Proceedings of the Institution of Mechanical Engineers, Part B: Journal of Engineering Manufacture. 1996;**210**(6):579-588

[33] Yang YH, Liu CH, Liang YH, et al. Hollow mesoporous hydroxyapatite nanoparticles (hmHANPs) with enhanced drug loading and pH-responsive release properties for intracellular drug delivery. Journal of Materials Chemistry B. 2013;**1**(19): 2447-2450. DOI: 10.1039/c3tb20365d

[34] Deng H, Lei Z. Preparation and characterization of hollow Fe_3O_4/SiO_2@ PEG–PLA nanoparticles for drug delivery. Composites Part B: Engineering. 2013;**54**:194-199. DOI: 10.1016/j.compositesb.2013.05.010

[35] Yang M, Li CH, Zhang YB, et al. Theoretical analysis and experimental research on temperature field of microscale bone grinding under nanoparticle jet mist cooling. Journal of Mechanical Engineering. 2018;**54**(18): 194-203. DOI: 10.3901/JME.2018.18.194

[36] Yang M, Li CH, Zhang YB, et al. Thermodynamic mechanism of Nanofluid minimum quantity lubrication cooling grinding and temperature field models. In: Mohsen Sheikholeslami Kandelousi, Microfluidics and Nanofluidics. London: IntechOpen; 2018. pp. 61-81

[37] Zhang YB, Li CH, Jia DZ, et al. Experimental evaluation of the lubrication performance of MoS_2/CNT nanofluid for minimal quantity lubrication in Ni-based alloy grinding. International Journal of Machine Tools and Manufacture. 2015;**99**:19-33. DOI: 10.1016/j.ijmachtools.2015.09.003

[38] Gao T, Li CH, Zhang YB, et al. Dispersing mechanism and tribological performance of vegetable oil-based CNT nanofluids with different surfactants. Tribology International. 2019;**131**:51-63. DOI: 10.1016/j.triboint.2018.10.025

[39] Zhang YB, Li CH, Jia DZ, et al. Experimental evaluation of MoS_2 nanoparticles in jet MQL grinding with different types of vegetable oil as base oil. Journal of Cleaner Production. 2015; **87**:930-940. DOI: 10.1016/j.jclepro.2014.10.027

[40] Lee PH, Nam JS, Li C, et al. An experimental study on micro-grinding process with nanofluid minimum quantity lubrication (MQL). International Journal of Precision Engineering and Manufacturing. 2012; **13**:331-338. DOI: 10.1007/s12541-012-0042-2

[41] Yang M, Li CH, Zhang YB, et al. Research on microscale skull grinding temperature field under different cooling conditions. Applied Thermal Engineering. 2017;**126**:525-537. DOI: 10.1016/j.applthermaleng.2017.07.183

Section 2

Energy Applications

Chapter 6

Material Removal Mechanism and Force Model of Nanofluid Minimum Quantity Lubrication Grinding

Yanbin Zhang, Changhe Li, Yongjun Zhao, Xin Cui, Xiufang Bai, Mingzheng Liu, Yali Hou, Min Yang, Naiqing Zhang, Heju Ji, Xiaoming Wang and Teng Gao

Abstract

Numerous researchers have developed theoretical and experimental approaches to force prediction in surface grinding under dry conditions. Nevertheless, the combined effect of material removal and plastic stacking on grinding force model has not been investigated. In addition, predominant lubricating conditions, such as flood, minimum quantity lubrication (MQL), and nanofluid minimum quantity lubrication (NMQL), have not been considered in existing force models. In this study, material removal mechanism under different lubricating conditions was researched. An improved theoretical force model that considers material removal and plastic stacking mechanisms was presented. Grain states, including cutting and ploughing, are determined by cutting efficiency (β). The influence of lubricating conditions was also considered in the proposed force model. Simulation was performed to obtain the cutting depth (a_g) of each "dynamic active grain." Parameter β was introduced to represent the plastic stacking rate and determine the force algorithms of each grain. The aggregate force was derived through the synthesis of each single-grain force. Finally, pilot experiments were conducted to test the theoretical model. Findings show that the model's predictions were consistent with the experimental results, with average errors of 4.19% and 4.31% for the normal and tangential force components, respectively.

Keywords: grinding, minimum quantity lubrication, nanofluid, material removal mechanism, force model

1. Introduction

With need of green manufacturing, nanofluid minimum quantity lubrication (NMQL) was introduced to grinding [1, 2], which could effectively improve cooling and lubricating performance compared to conventional dry or flood conditions [3, 4]. Furthermore, with the deeper understanding of grinding mechanism and the development of process technology for difficult-to-grinding material, advancement in the experiment has been rapidly made [5–7]. Of course, with such a demand from designing of grinding technological parameter, material removal mechanism and force model under NMQL condition are of key importance.

Material removal mechanism, here, mostly refers to deformation, fracture process, and strain rate of material in grinding zone, as well as mechanical behavior of chips and furrow formation [8, 9], which relates to grinding parameters and cooling/lubricating conditions [10, 11]. For dry grinding, the experimental studies have been carried out, and force, temperature, shape of chips, and surface quality were observed, which showed special phenomenon when higher speed (i.e., speed effect) and lower cutting depth (i.e., size effect) were used. Nevertheless, material removal mechanism under NMQL condition is vacant and important, which should be researched deeply.

Force model of grinding, in general, was vector sum of sliding force, cutting force, and ploughing force of all dynamic active grains in grinding zone. Many researches were carried out to establish grinding force model under dry condition. Werner [12] derived a grinding force equation with two parts (cutting force and sliding force), which was adopted by Malkin and Hwang [13], Li et al. [14], Younis et al. [15], and Tang et al. [16]. To obtain grain stages and numbers, Rayleigh's probability density distribution [17, 18] and normal distribution [19] were used to describe cutting depth of grain [20]. Except for steel material, the grinding force model was also researched for single crystal sapphire [21], SiCp/Al composites [22], complex optical mirrors [23], etc. However, previous models were limited at three aspects:

i. The average cutting depths were used for each grain states (i.e., cutting, ploughing, sliding), which cannot represent the actual situation in the grinding zone [24].

ii. The critical depth of cutting and ploughing was determined by an experience-based evaluation method: "when the grain-cutting depth reached 0.05 times of the radius of cutting grains, cutting action occurred" [25–27]. It's imprecise for different material.

iii. The number of "static active grains" was presented to develop force models, which are not considered interference effect of grains on each other [28–32].

iv. The lubricating condition was not considered.

This chapter proposes an improved theoretical force model that takes the material removal and plastic stacking mechanism into consideration. The scratch tests reveal the relationship between cutting depth (a_g) and cutting efficiency (β) to distinguish the cutting and ploughing grains and determine the force algorithms of each grain. The tribological tests reveal the friction coefficients under different grinding lubricating conditions to calculate the frictional force of each grain. In addition, the a_g of each "dynamic active grain" in the grinding zone is obtained by simulation, to assist the development of the grinding force model. To verify the proposed model, grinding experiments were performed, and force values were obtained and compared with predictive values.

2. Material removal mechanism of NMQL grinding

2.1 Deformation and strain rate

The material deformation mechanism in the grinding zone with single abrasive particle is shown in **Figure 1**. The material deformation mechanism in the chip

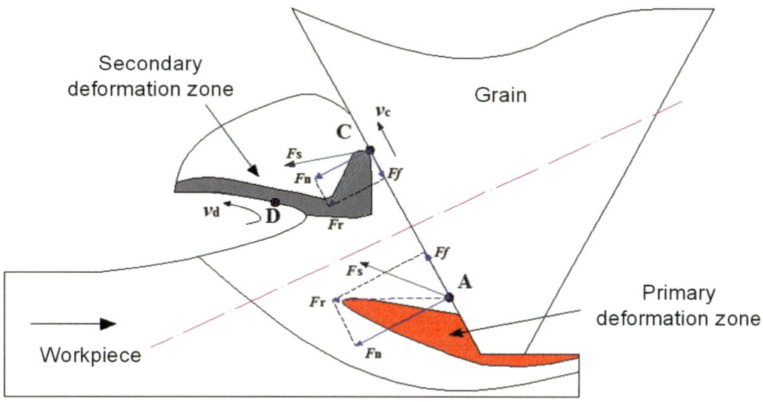

Figure 1.
Deformation mechanism of material in the cutting zone.

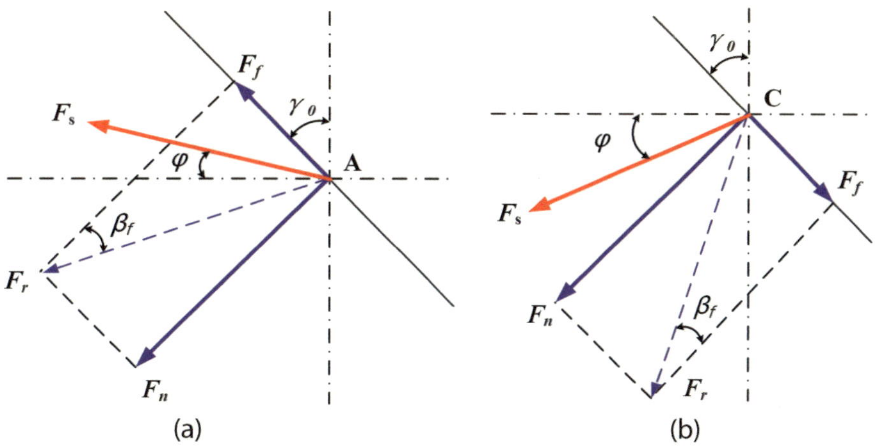

Figure 2.
Stress analysis of the workpiece material in the cutting zone. (a) Primary deformation zone and (b) secondary deformation zone.

formation zone during the grinding machining is identical with that during cutting machining, and primary deformation zone and secondary deformation zone simultaneously exist.

Ding et al. [33] conducted an analog simulation of high-temperature cutting of nickel-based alloy workpiece material with single abrasive particle and also observed the abovementioned phenomenon. Different from cutting machining, the negative rake cutting form of abrasive particle leads to a different mechanical action mechanism between abrasive particle/grinding chips; the calculation formulas of shear angle φ, frictional angle β_f, and strain rate are also changed, but the impact trends of grinding process parameters and lubricating conditions on material removal mechanism are totally the same. The force analysis of the grinding zone of single abrasive particle is shown in **Figure 2**, and material force bearing and deformation mechanism can be described as follows:

i. In the primary deformation zone, **Figure 2(a)** is force analysis graph of point A, and the resultant force of frictional force F_f and extrusion force F_n is F_r. According to the mechanical theory of material, the included angle between rupture chip formation direction (direction F_s of shear zone) and resultant force F_r is $\pi/4$, and shear angle φ is on horizontal line; at the time, frictional force F_f

exerts a promoting effect on rupture chip formation of the material upward along the abrasive particle/cutting chip interface. Hence, the expression of shear angle φ is solved according to geometrical relations of several angles as below:

$$\varphi_1 = \frac{\pi}{4} + \beta_f - \gamma_0 \tag{1}$$

The ratio of thickness to length of grinding chips formed through cutting machining of abrasive particle is about 1/100, so negative rake is $\gamma_0 \approx \theta/2$ and frictional angle is β_f = arctan(μ) under abrasive particle cutting, where μ is frictional coefficient between rake face of abrasive particle/cutting chips. The relational expression between γ_0 and β is substituted into Eq. (1) to obtain:

$$\varphi_1 = \frac{\pi}{4} + \arctan\mu - \frac{\theta}{2} \tag{2}$$

ii. In the secondary deformation zone, **Figure 2(b)** is force analysis chart, where the resultant force of frictional force F_f and extrusion force F_n is F_r. According to the mechanical theory of material, the included angle between rupture chip formation direction (direction F_s of shear zone) and resultant force F_r is $\pi/4$, and shear angle φ is below the horizontal line; at the time, frictional force F_f exerts an inhibitory effect on flow of grinding chips downward along the abrasive particle/cutting chip interface. After kinetic analysis, speed of movement on abrasive particle/grinding chip interface is low due to frictional force, but the speed on free face of grinding chips is high; under high-speed cutting condition, transient speed difference is generated between abrasive particle/grinding chip interface and free face of grinding chips due to high strain rate of grinding chips, thus generating a shear zone and forming the secondary deformation zone. Hence, the expression of shear angle φ can be solved according to geometrical relations of several angles as:

$$\phi_2 = \beta_f + \gamma_0 - \frac{\pi}{4} = \arctan\mu + \frac{\theta}{2} - \frac{\pi}{4} \tag{3}$$

For high-speed grinding machining, the change of strain rate is the primary factor influencing the material removal mechanism. The strain rate reaches as high as 10^4–$10^5\ s^{-1}$ in the cutting machining process, but for low-speed grinding machining ($V_s \leq$ 30 m/s), the strain rate of material removal reaches up to 10^7–$10^8\ s^{-1}$ because of negative rake characteristic of abrasive particle. Therefore, it's speculated that the strain rate under high-speed grinding machining condition is higher than that under low-speed grinding machining condition of grinding wheel by 1 or 2 orders of magnitudes. Under high-speed grinding machining condition, the strain hardening effect and strain strengthening effect caused by strain rate of material deformation generate a remarkable effect on the material removal mechanism. Jin and Stephenson [34] established the equation of shear strain in the shear zone of abrasive particle and strain rate. Taking maximum undeformed cutting thickness and shear angle in the primary deformation zone and secondary deformation zone into account, strain rate of grinding chips in the two deformation zones could be calculated by the following equations:

$$\dot{\gamma}_1 = \frac{\lambda v_s^2 \cos(\frac{\theta}{2}) \sin(\frac{\pi}{4} + \arctan\mu - \frac{\theta}{2})}{\pi v_w \sqrt{a_p D} \cos(\frac{\pi}{4} + \arctan\mu)} \tag{4}$$

$$\dot{\gamma}_2 = -\frac{\lambda v_s^2 \cos(\frac{\theta}{2}) \sin(\arctan\mu + \frac{\theta}{2} - \frac{\pi}{4})}{\pi v_w \sqrt{a_p D} \cos(\arctan\mu + \theta - \frac{\pi}{4})} \tag{5}$$

where φ is shear angle, θ is vertex angle of abrasive particle, λ_1 is average length-width ratio in the shear zone (λ_1 = 6–12), v is cutting speed of abrasive particle, and $v \approx V_s$ is for single abrasive particle grinding.

It can be known from Eq. (4) that factors influencing deformational strain rate of workpiece material during the cutting process with single abrasive particle include grinding parameters, abrasive particle shape, and lubrication characteristics of abrasive particle/grinding chip interface. In this study, as $V_s/V_w = 1 \times 10^4$ and abrasive particle shape are unchanged, the influence trends of grinding speed V_s, grinding depth V_w, and grinding coefficient μ on shear rate will be hereby discussed.

As shown in **Figure 3(a)**, when other parameters remain unchanged, the strain rates in both primary shear zone and secondary shear zone present a linear growth trend when V_s increases from 30 to 120. In comparison with the primary deformation zone, the strain rate in the secondary deformation zone has a higher growth rate.

According to **Figure 3(b)**, when the frictional coefficient μ gradually increases, the strain rate in the primary deformation zone presents an increasing trend of quadratic function, because the frictional force in the primary deformation zone exerts a gaining effect on material removal and the strain rate increases with frictional force, which is better for chip formation. In the secondary deformation zone, as the frictional coefficient μ increases from small to great (characterizing lubricating condition), the strain

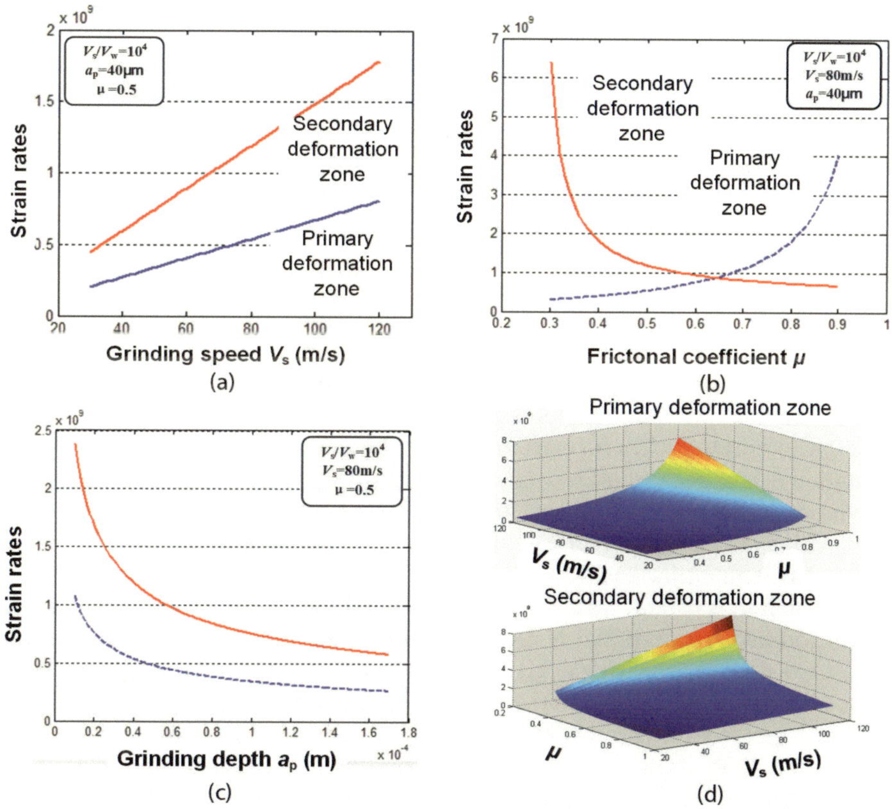

Figure 3.
Strain rate under different grinding parameters and frictional coefficient. (a) Influence of grinding speed, (b) influence of frictional coefficient, (c) influence of grinding depth and (d) two-factor analysis.

rate in the secondary deformation zone presents a declining trend of quadratic function, the frictional force in the primary deformation zone prevents grinding chips from flowing out, and flow direction of grinding chips is gradually turned from towards abrasive particle into along the surface of abrasive particle, so the strain rate gradually declines. As shown in **Figure 3(c)**, the strain rate presents a linear declining trend due to increase of grinding depth. As this study focuses on speed effect and lubricating effect in the high-speed grinding machining, in consideration of impact trends of grinding speed and frictional coefficient on strain rate, **Figure 3(d)** shows the variation trend graphs of the primary deformation zone and secondary deformation zone under the influence of grinding speed and frictional coefficient.

2.2 Mechanism of chip formation

2.2.1 Theoretical research

Impact dynamics describes material deformation behaviors under high speed as an adiabatic shear process. Adiabatic shear effect refers to constitutive instability (thermal viscoplastic instability) of the material under impact load. Extremely strain rate exists in the material removal shear zone under the action of impact load, and the shear zone is a thermal insulating environment within a very short time. At the time, nonelastic energy in the material deformation process is converted into a large quantity of grinding heat, which leads to abrupt temperature rise of the material in the shear zone and declination of material hardness (softening) under temperature rise. The softening effect exceeds strain hardening effect and strain strengthening effect, and thus shear zone is formed and material instability is caused. Hence, the chip formation process under the effect of abrasive particle is dynamic stress and thermal force coupling action process under a high strain rate.

The removal process of metal material derives from plastic deformation after the material strain increases. In the previous studies, scholars have conducted a large quantity of dynamic material tensile tests [35], proving strain hardening and strain strengthening phenomena. As shown in **Figure 4**, the material removal process with the change of strain rate can be divided into the three following types according to the stress-strain curve:

As shown in **Figure 5(a)**, under quasi-static conditions, the material deformation stress will increase obviously with the strain, indicating that notable strain hardening effect happens to the workpiece material. Grinding chips will be formed after the stress increases to material breaking limit σb1, and free face of cutting chips presents periodic upheaval (bamboo shape), which is caused by extruding deformation in the grinding chip formation process; however, the strain rate of material deformation is low because of low speed, and no obvious shear zone is formed.

As shown in **Figure 5(b)**, under high strain rate, assume that the shear zone of grinding chips is in an isothermal environment: the material deformation stress in the strain hardening phase is far higher than that under quasi-static conditions, and this phenomenon is called "strain strengthening effect" of plastic deformation of material, and its deformational resistance is enhanced under strain strengthening effect. At the time, as abrasive particle has a strong impact effect on the material, workpiece material forms shear layers under high strain rate so as to form grinding chips, and rupture stress limit between shear layers is breaking limit σ_{b2} of the material under high strain rate. Free face of grinding chips presents notable periodic shear slippage layer, and the slippage distance of shear layer under high strain condition is obviously higher than that under quasi-static state; and the higher the speed, the higher the strain rate, and the greater the slippage distance of shear layer; and even radical rupture and separation between shear layers is caused by bluff-type rupture process.

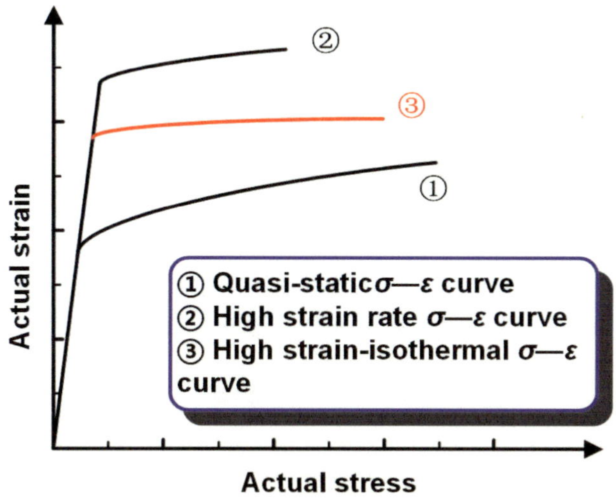

Figure 4.
Stress-strain curve under "velocity effect."

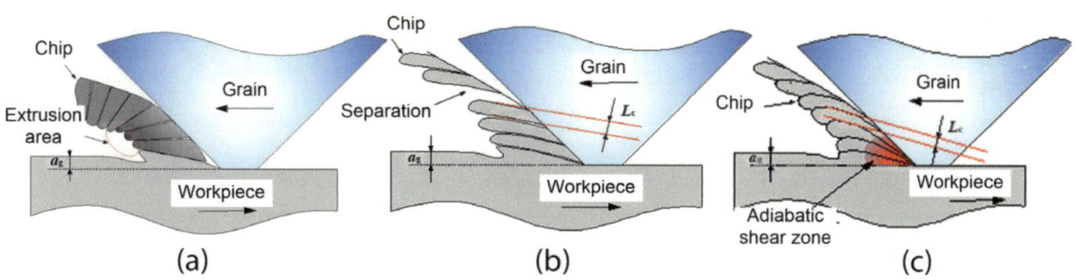

Figure 5.
Chip-forming mechanism under (a) quasi-static, (b) high strain isothermal, and (c) high strain condition.

As shown in **Figure 5(c)**, in consideration of temperature rise at abrasive particle/grinding chip interface under high strain rate, the plastic deformation process of the material at the time is a comprehensive result of strain hardening effect and thermal softening effect, namely, adiabatic shear process. In **Figure 4**, the stress-strain curve is between quasi-static curve (curve 1) and high strain isothermal condition (curve 2) under the action of adiabatic shear, because a large quantity of grinding heat softens the material so as to reduce deformational resistance under the action of strengthening strain rate. As the strain rate increases, the heat transferred to shear zone is also increasing, thermal softening effect becomes more obvious, and the breaking limit relation of the material under three boundary conditions is $\sigma_{b2} > \sigma_{b3} > \sigma_{b1}$. Material rupture and plastic flow after thermal softening simultaneously exist between shear layers of grinding chips.

Hence, the critical research hotpots regarding material removal mechanism under high-speed grinding conditions are to explore into influences of grinding parameters and lubricating conditions on strain strengthening effect and thermal softening effect and build an evaluation model.

2.2.2 Experimental research

SEM of chips were shown in **Figure 6**.

The grinding chip shapes are similar at the grinding speed of 40 m/s under three different working conditions, namely, continuous banding chips, because the material strain rate is low and thermal softening effect in the grinding zone is small under low grinding speed. When the grinding speed is 80 m/s, obvious shear zones

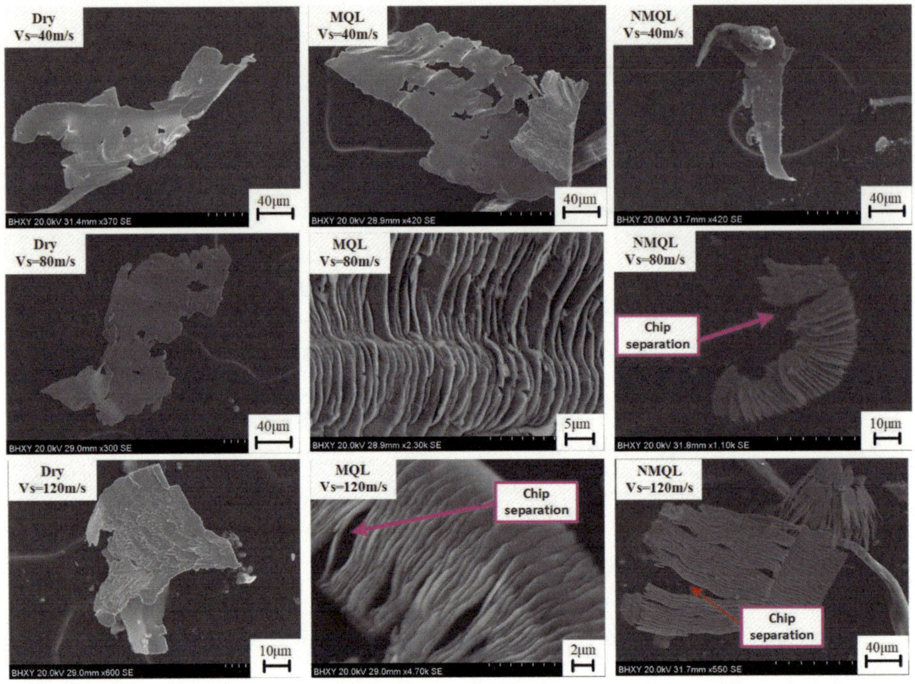

Figure 6.
Chip morphology at different lubricating conditions.

are observed on the free faces of the three types of grinding chips, indicating that under high grinding speed, the material experiences adiabatic shear during the forming process. Differently, the shear slippage layer spacing obtained under dry grinding conditions is small, that under MQL condition is larger, and that under NMQL is the largest, where even rupture takes place.

On the one hand, as the lubricating effect is enhanced and frictional coefficient is reduced in succession under the three lubricating conditions, the strain rate in the primary deformation zone at the same speed increases successively, so does shear slippage distance under the three conditions. On the other hand, the output of frictional heat is reduced as the lubricating effect is enhanced successively under the three conditions, and the heat quantity transferred out of the shear zone is increased as the cooling effect is enhanced successively under the three conditions. Therefore, the thermal softening effect during NMQL cutting chip formation process is lower than that under dry grinding condition. The shear layer slippage of grinding chips is mainly material rupture, followed by plastic flow, so shear layer spacing is larger, and grinding chips even rupture. Under the grinding speed of 120 m/s, the abovementioned rule is magnified again. Banding chips are still formed due to thermal softening effect under dry grinding, rupture of grinding chips induced by excessive shear layer spacing also appears under MQL condition, and the rupture under NMQL condition is further aggravated.

2.3 Mechanism of furrow formation and cutting efficiency

2.3.1 Plastic stacking effect

Cutting depth a_g (varies from 0 to a_{gmax}) and critical value a_{gc} were the key points for distinguishing ploughing and cutting stage. The value a_{gc} could be gotten from cutting experiment of single grain for different materials. The change rule could be concluded from **Figure 7**.

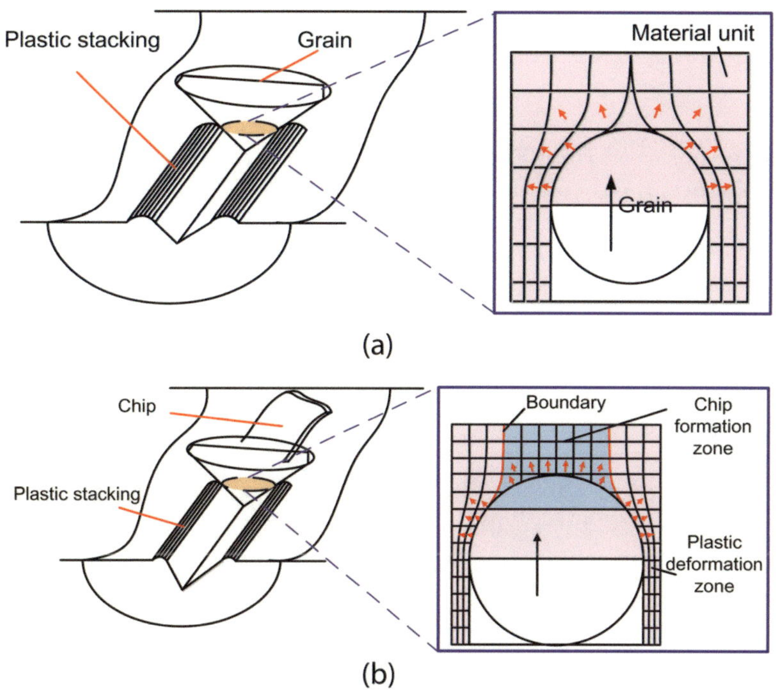

Figure 7.
Material deformation behavior in cutting of a single grain. (a) Ploughing grain and (b) cutting grain.

i. When $a_g \le a_{gc}$ (ploughing stage). Material of workpiece flows upheaved and stacked at two sides of the furrows because the stress between material units is lower than fracture stress.

ii. When $a_g > a_{gc}$ (cutting stage). Plastic stacking theory [36], which describes the coexistence phenomenon of material removal and elastic-plastic flow in grinding, is a good method to understand deformation behavior of material. Pink-colored material units in **Figure 7(b)** were the elastic-plastic flow part, and blue-colored units were the material removal part. However, a boundary was born between two regions.

Therefore, chip formation region (0–α_1) and elastic-plastic flow region (α_1–π/2) were defined in this paper, in which two were differentiated by critical value of angle α_1, as shown in **Figure 8**. α_1 (could be calculated by cutting efficiency β [37]) was different when material changes:

$$\alpha_1 = \arccos(\sqrt{\beta}) \tag{6}$$

2.3.2 Model of $\beta(\alpha_g)$

Parameter β will increase from 0 to almost 1 when a_g increases [38]. Therefore the model of $\beta(\alpha_g)$ is the key point of grinding force prediction. Cutting experiment of single grain was carried out for stainless steel material (440 C, Ra = 0.04–0.05 μm). Surface topography of furrow is shown in **Figure 9**, and numerical fitting of $\beta(\alpha_g)$ was shown in **Figure 10**.

The $\beta(\alpha_g)$ curve present S-shaped (when a_g varies from 0 to 4.5 μm) and further linear trend (when ag ≥ 4.5 μm) . Therefore, Gompertz growth equation and Gaussian fitting were considered simultaneously, and piecewise function (R^2 = 0.9945) is shown as Eq. (7).

$$\beta(a_g) = \begin{cases} a_1 \cdot e^{(-(\frac{a_g - b_1}{c_1})^2)} + a_2 \cdot e^{(-(\frac{a_g - b_2}{c_2})^2)} + \cdots + a_6 \cdot e^{(-(\frac{a_g - b_6}{c_6})^2)}, 0 \le x \le 3.8\mu m \\ k \cdot e^{(-a \cdot e^{-b \cdot a_g})}, x > 3.8\mu m \end{cases} \tag{7}$$

It could be observed from **Figure 10** that (i) three stages could be learned from curve slope of $\beta(\alpha_g)$; (ii) compared to study of Hahn [39], sliding stage (0–0.023 μm) is declined; and (iii) the transition stage exists between ploughing and cutting stage, which is different from traditional theory.

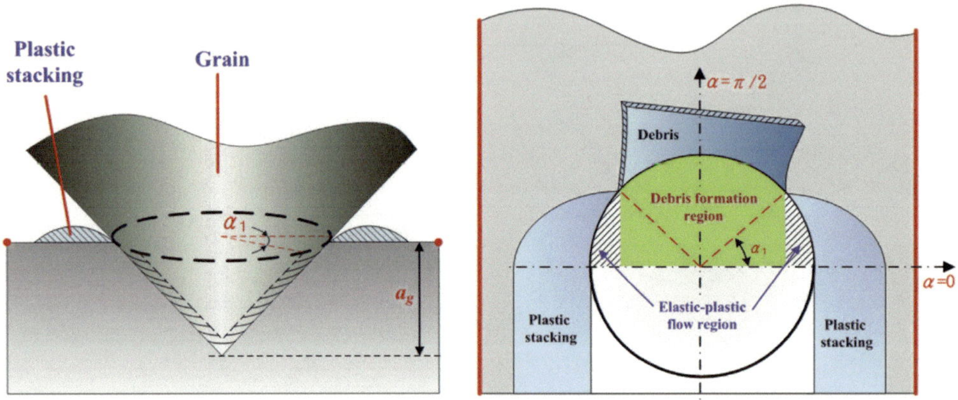

Figure 8.
Plastic stacking mechanism.

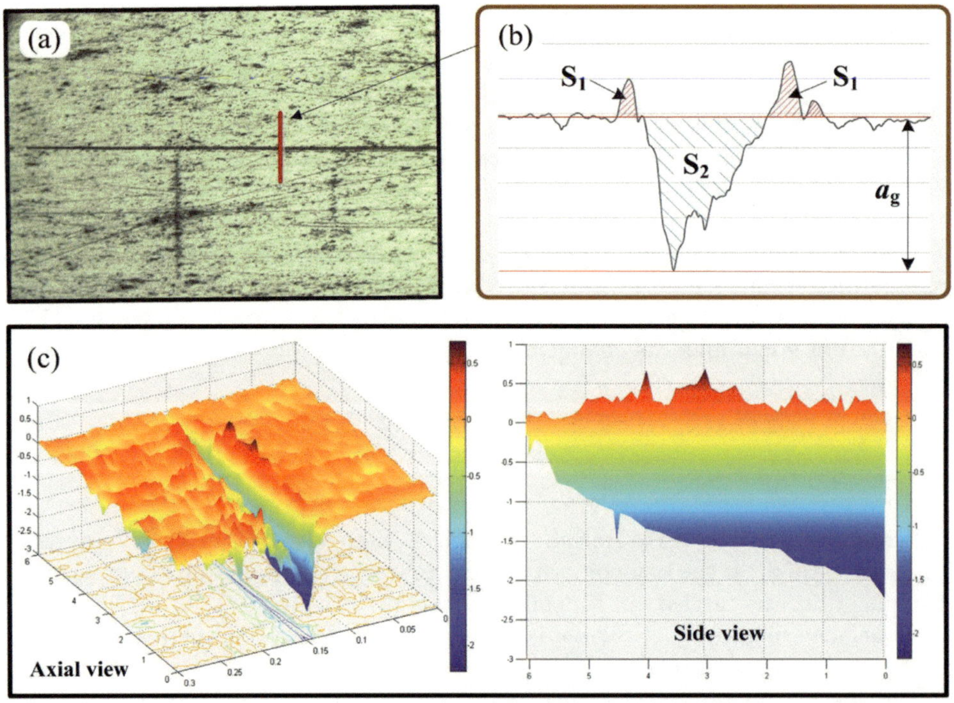

Figure 9.
Result of scratch tests. (a) Surface topography of furrow, (b) 2D profile curve of furrow, (c) 3D graph of furrow.

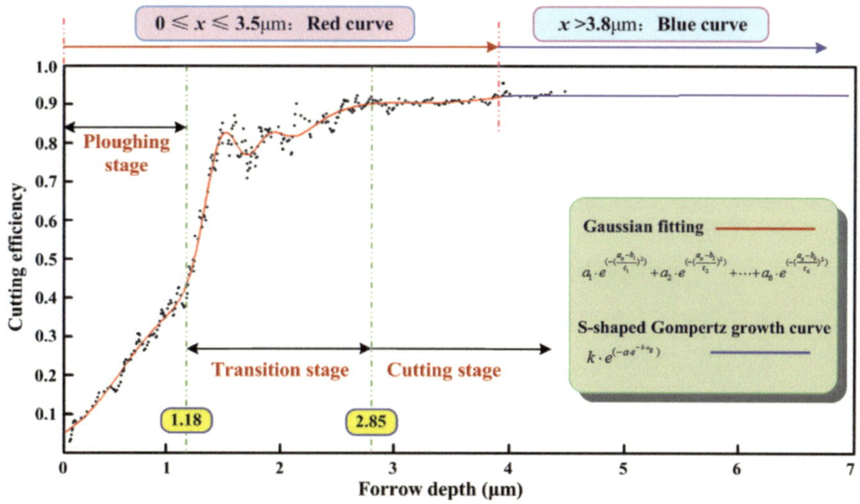

Figure 10.
Equation of cutting efficiency (β).

3. Force model of single grain

3.1 Stress state of grain

As a premise for force model, analysis of stress state was carried out in this section according to the above results, as shown in **Figure 11**. Plastic flow is the material behavior, and equal stress value (yields stress δ_s) distributes on grain surface. The stresses at chip formation region (δ_0) and elastic-plastic flow region (δ_1) show characteristics as follows: (i) equal stress value (fracture stress δ_b) distributes on chip formation region and (ii) linear formula of stress distributes on elastic-plastic flow region, which varies from 0 to δ_b.

3.2 Force model in chip formation region (α_1–$\pi/2$)

δ_0 was composed of δ_{01} and δ_{02} in chip formation region. As shown in **Figure 12**, the integral unit (ds) can be expressed as:

$$d_s = \frac{a_g^{\ 2} \cdot \tan\theta}{2 \cdot \cos\theta} \cdot d\alpha \tag{8}$$

The plastic flow force equation $F_{tc(01)}(a_g)/F_{nc(01)}(a_g)$ in chip formation region could be calculated as:

$$F_{tc(01)}(a_g) = \int_{\alpha_1}^{\frac{\pi}{2}} \delta_s \cdot a_g^{\ 2} \cdot \tan\theta \cdot \cos\alpha \cdot d\alpha \tag{9}$$

$$F_{nc(01)}(a_g) = \int_{\alpha_1}^{\frac{\pi}{2}} \delta_s \cdot a_g^{\ 2} \cdot \tan^2\theta \cdot d\alpha = (\frac{\pi}{2} - \alpha_1) \cdot \delta_s \cdot a_g^{\ 2} \cdot \tan^2\theta \tag{10}$$

where θ is grain vertex angle and α_1 could be obtained by Eqs. (6) and (7), respectively.

The chip formation force equation $F_{tc(02)}(a_g)$ in chip formation region could be calculated as:

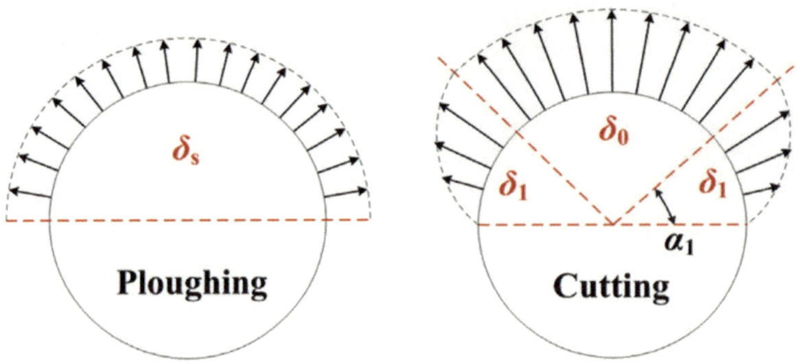

Figure 11.
Schematic diagram of the grain stress state.

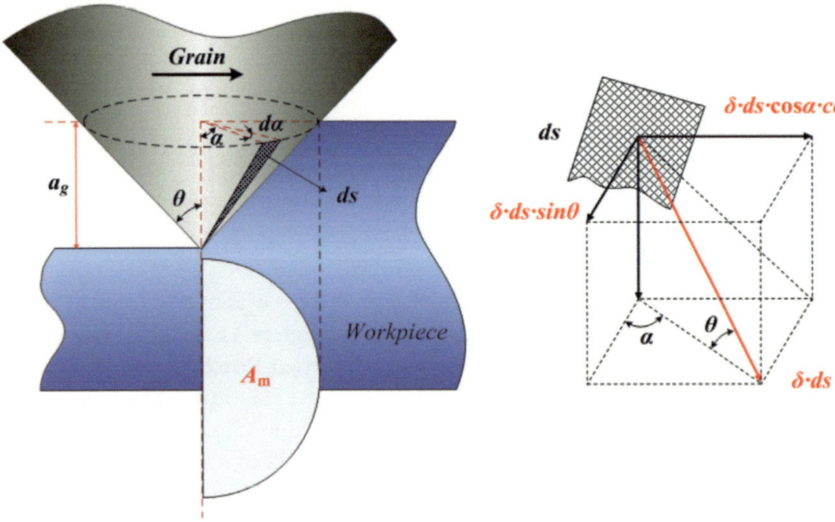

Figure 12.
Calculation principle of cutting force.

$$F_{tc(02)}(a_g) = \int_{\alpha_1}^{\frac{\pi}{2}} \delta_{02} \cdot a_g{}^2 \cdot \tan\theta \cdot \cos\alpha \cdot d\alpha = \delta_b \cdot A_m \tag{11}$$

where A_m is area of the chip formation region and δ_{02} could be obtained by Eq. (11):

$$\delta_{02} = \frac{\pi \cdot \tan\theta}{2 \cdot (1 - \sin\alpha_1)} \cdot \delta_b \tag{12}$$

The normal force could be further calculated by Eqs. (11) and (12):

$$F_{nc(02)}(a_g) = \int_{\alpha_1}^{\frac{\pi}{2}} \delta_{02} \cdot a_g{}^2 \cdot \tan^2\theta \cdot d\alpha = \frac{\pi \cdot (\frac{\pi}{2} - \alpha_1)}{2 \cdot (1 - \sin\alpha_1)} \cdot \delta_b \cdot a_g{}^2 \cdot \tan^3\theta \tag{13}$$

In summary, δ_0 can be expressed as:

$$\delta_0 = \delta_{01} + \delta_{02} = \delta_s + \frac{\pi \cdot \tan\theta}{2 \cdot (1 - \sin\alpha_1)} \cdot \delta_b \tag{14}$$

3.3 Force model in elastic-plastic flow region (0– α_1)

The equation of δ_1 and force in elastic-plastic flow region show increased trend and could be calculated as:

$$\delta_1(\alpha)=\left[\frac{\delta_s}{\alpha_1}+\frac{\pi\cdot\delta_b\cdot\tan\theta}{\alpha_1\cdot(1-\sin\alpha_1)}\right]\cdot\alpha \tag{15}$$

$$F_{tc(1)}(a_g)=\int_0^{\alpha_1}\delta_1(\alpha)\cdot a_g{}^2\cdot\tan\theta\cdot\cos\alpha\cdot d\alpha \tag{16}$$

$$F_{nc(1)}(a_g)=\int_0^{\alpha_1}\delta_1(\alpha)\cdot a_g{}^2\cdot\tan^2\theta\cdot d\alpha \tag{17}$$

Force equation of cutting grain could be obtained by the above equation:

$$\begin{aligned}F_{tc}(a_g)&=F_{tc(1)}(a_g)+F_{tc(01)}(a_g)+F_{tc(02)}(a_g)\\&=\int_0^{\alpha_1}\delta_1(\alpha)\cdot a_g{}^2\cdot\tan\theta\cdot\cos\alpha\cdot d\alpha+\int_{\alpha_1}^{\frac{\pi}{2}}\delta_s\cdot a_g{}^2\cdot\tan\theta\cdot\cos\alpha\cdot d\alpha+\delta_b\cdot A_m\end{aligned} \tag{18}$$

$$\begin{aligned}F_{nc}(a_g)&=F_{nc(1)}(a_g)+F_{nc(01)}(a_g)+F_{nc(02)}(a_g)\\&=\left[\int_0^{\alpha_1}\delta_1(\alpha)\cdot d\alpha+\frac{\pi\cdot(\frac{\pi}{2}-\alpha_1)}{2\cdot(1-\sin\alpha_1)}\cdot\delta_b\cdot\tan\theta+(\frac{\pi}{2}-\alpha_1)\cdot\delta_s\right]\cdot a_g{}^2\cdot\tan^2\theta\end{aligned} \tag{19}$$

3.4 Force model of cutting and ploughing grain

Material behavior in ploughing stage could be categorized as plastic flow, in which stress should reach at δ_s. Therefore, force model of ploughing grain could be calculated as Eqs. (20) and (21).

$$F_{tp}(a_g)=\int_0^{\frac{\pi}{2}}\delta_s\cdot a_g{}^2\cdot\tan\theta\cdot\cos\alpha\cdot d\alpha \tag{20}$$

$$F_{np}(a_g)=\int_0^{\frac{\pi}{2}}\delta_s\cdot a_g{}^2\cdot\tan^2\theta\cdot d\alpha=\frac{\pi}{2}\cdot\delta_s\cdot a_g{}^2\cdot\tan^2\theta \tag{21}$$

3.5 Frictional force model

The frictional force is composed of rake face on ploughing grains $F_{pf}(a_g)$, rake face on cutting grains $F_{cf}(a_g)$, and the wear plane of the grain (f_n, f_t), as shown in Eqs. (22)–(30). The force magnitude is determined by stress and the lubricating state between the grains and the workpiece.

$$F_{pf}(a_g)=2\cdot\int_0^{\frac{\pi}{2}}\mu\cdot\delta_s\cdot ds=\int_0^{\frac{\pi}{2}}\mu\cdot\delta_s\cdot a_g{}^2\cdot\frac{\tan\theta}{\cos\theta}\cdot d\alpha \tag{22}$$

$$F_{tpf}(a_g)=F_{pf}(a_g)\cdot\sin\theta=\int_0^{\frac{\pi}{2}}\mu\cdot\delta_s\cdot a_g{}^2\cdot\tan^2\theta\cdot d\alpha \tag{23}$$

$$F_{npf}(a_g)=F_{pf}(a_g)\cdot\cos\theta=\int_0^{\frac{\pi}{2}}\mu\cdot\delta_s\cdot a_g{}^2\cdot\tan\theta\cdot d\alpha \tag{24}$$

$$\begin{aligned}F_{cf}(a_g)&=2\cdot(\int_{\alpha_1}^{\frac{\pi}{2}}\mu\cdot\delta_0\cdot ds+\int_0^{\alpha_1}\mu\cdot\delta_1\cdot ds)\\&=\int_{\alpha_1}^{\frac{\pi}{2}}\mu\cdot\delta_0\cdot a_g{}^2\cdot\frac{\tan\theta}{\cos\theta}\cdot da+\int_0^{\alpha_1}\mu\cdot\delta_1\cdot a_g{}^2\cdot\frac{\tan\theta}{\cos\theta}\cdot da\end{aligned} \tag{25}$$

$$F_{tcf}(a_g) = F_{cf}(a_g) \cdot \sin\theta = \int_{\alpha_1}^{\frac{\pi}{2}} \mu \cdot \delta_0 \cdot a_g^{\ 2} \cdot \frac{\tan\theta}{\cos\theta} \cdot da + \int_0^{\alpha_1} \mu \cdot \delta_1 \cdot a_g^{\ 2} \cdot \frac{\tan\theta}{\cos\theta} \cdot da \tag{26}$$

$$F_{ncf}(a_g) = F_{cf}(a_g) \cdot \cos\theta = \int_{\alpha_1}^{\frac{\pi}{2}} \mu \cdot \delta_0 \cdot a_g^{\ 2} \cdot \tan\theta \cdot da + \int_0^{\alpha_1} \mu \cdot \delta_1 \cdot a_g^{\ 2} \cdot \tan\theta \cdot da \tag{27}$$

$$f_n = N_d \cdot S_w \cdot \bar{p} = \frac{4 \cdot P_0 \cdot S_w \cdot N_d \cdot V_w}{V_s \cdot D} = \frac{4 \cdot K_1 \cdot N_d \cdot V_w}{V_s \cdot D} \tag{28}$$

$$f_t = \mu \cdot N_d \cdot S_w \cdot \bar{p} = \frac{4 \cdot \mu \cdot P_0 \cdot \delta \cdot N_d \cdot V_w}{V_s \cdot D} = \frac{4 \cdot \mu \cdot K_1 \cdot N_d \cdot V_w}{V_s \cdot D} \tag{29}$$

$$K_1 = P_0 \cdot S_w \tag{30}$$

where K_1 is the physical quantity related to the grain's wear state of the grinding wheel, which could be reversely solved through grinding experiment, and μ is the friction coefficient between the workpiece and the grains, which could be obtained by tribological tests [40].

4. Grinding force model and prediction

4.1 Procedure of modeling common grinding wheel

The matrix of the location distribution of the grains in the grinding zone is established as $G(z_g)$. The matrix of the protrusion height of the grains can be calculated as [40]:

$$G(z) = \frac{G(d)}{2} + G(z_g) = \begin{bmatrix} z_{11} & z_{12} & \cdots & z_{1N_y} \\ z_{21} & z_{22} & \cdots & z_{2N_y} \\ \vdots & \vdots & \ddots & \vdots \\ z_{N_x1} & z_{N_x1} & \cdots & z_{N_xN_y} \end{bmatrix} \tag{31}$$

4.2 Dynamic active grains in grinding zone

For two continuous cutting grains, maximum undeformed chip thickness a_{gmax} ($a_g = 0.5a_{gmax}$ [42]) could be solved as [41, 42]:

$$a_{g\max} = 2 \cdot \lambda \cdot \frac{V_w}{V_s} \cdot \sqrt{\frac{a_p}{D}} \tag{32}$$

where λ is the space between the continuous cutting grains, D is the diameter of the grinding wheel, V_w is the feed speed, and V_s is the peripheral speed of the grinding wheel.

For common grinding wheel, λ value represented the space between the dynamic active grains. The equation of the nth grain is deduced as:

$$a_{g\max(n)} = 2 \cdot \lambda_{(n\sim n-1)} \cdot \frac{V_w}{V_s} \cdot \sqrt{\frac{a_p}{D}} + (a_{p(n)} - a_{p(n-1)}) \tag{33}$$

where $a_{gmax(n)}$ is the maximum undeformed chip thickness of the nth dynamic active grain, $\lambda_{(n\sim n-1)}$ is the space between the nth and the (n − 1)th dynamic active grains, $a_{p(n)}$ is the protrusion height of the nth dynamic active grain, and a_p(n − 1) is the protrusion height of the (n − 1)th dynamic active grain.

4.3 Grinding force model

Concrete calculation of grinding force can be described in **Figure 13**; the grinding force can be expressed in Eqs. (34) and (35).

$$F_t = \sum_1^{N_c}[F_{tc}(a_{gn}) + F_{tcf}(a_{gn})] + \sum_1^{N_p}[F_{tp}(a_{gm}) + F_{tpf}(a_{gm})] + f_t \tag{34}$$

$$F_n = \sum_1^{N_c}[F_{nc}(a_{gn}) - F_{ncf}(a_{gn})] + \sum_1^{N_p}[F_{np}(a_{gm}) - F_{npf}(a_{gm})] + f_n \tag{35}$$

where F_t/F_n are the tangential/normal grinding force, respectively, and f_t/f_n are the tangential/normal frictional force on the wear plane of grains, respectively. For the nth cutting grains ($1 \le n \le N_c$), a_{gn} is the cutting depth; $F_{tc}(a_{gn})/F_{nc}(a_{gn})$ are the tangential/normal cutting force, respectively; and $F_{tcf}(a_{gn})/F_{ncf}(a_{gn})$ are the tangential/normal frictional force on the rake face of the cutting grain, respectively. For the mth ploughing grains ($1 \le n \le N_c$, $N_c + N_p = N_d$), a_{gm} is the cutting depth; $F_{tp}(a_{gm})/F_{np}(a_{gm})$ are the tangential/normal, respectively; and $F_{tpf}(a_{gm})/ F_{npf}(a_{gm})$ are the tangential/normal frictional force on the rake face of ploughing grain, respectively.

4.4 Experimental verification

MoS_2-palm oil nanofluid minimum quantity lubrication grinding experiment was carried out with K-P36 surface grinder. Prediction and experimental force values was shown in **Figure 14**.

The average percentage of the deviation in the normal force is 4.19%, while in the tangential force is 4.31%. For a certain grinding condition (dry grinding, V_s = 20 m/s, V_w = 2 m/min, a_p = 15 μm), the contribution of the tangential frictional force in the total tangential grinding force is approximately 89.17% and approximately 90.71% for normal force. However, it was decreased to 86.52% and 89.43% for tangential and normal direction, respectively.

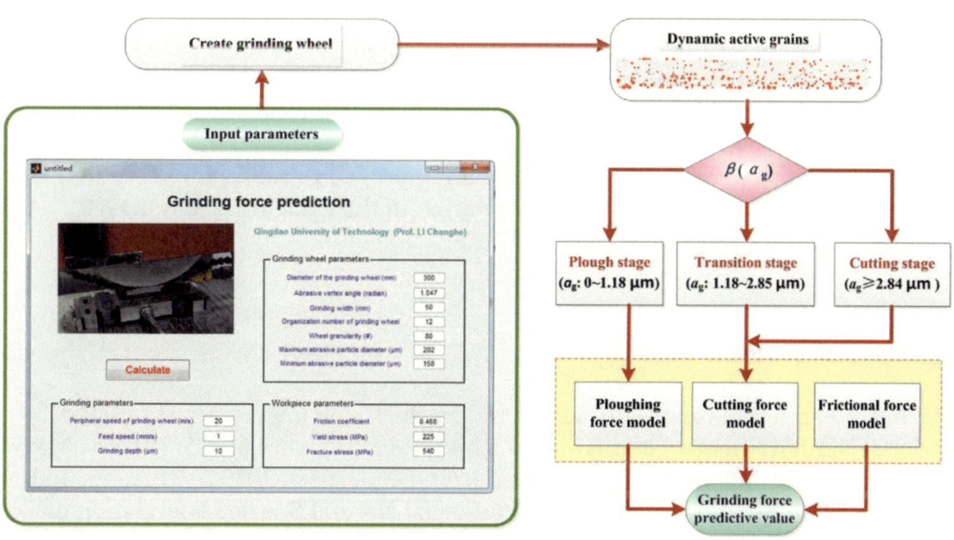

Figure 13.
Process of the grinding force predictive program.

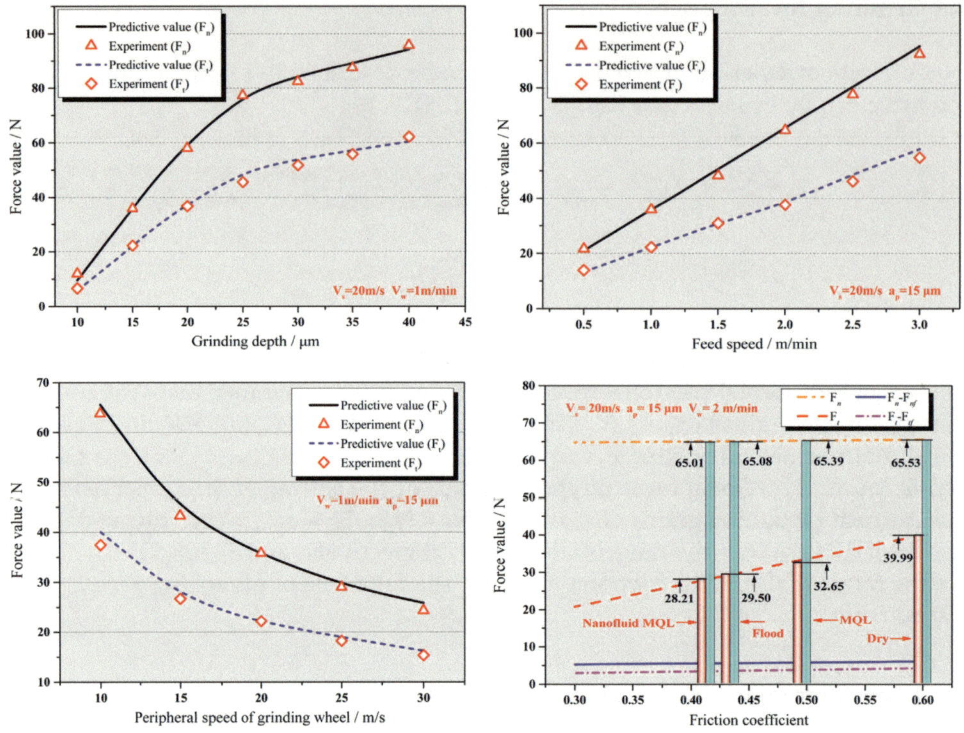

Figure 14.
Force prediction and experiment result.

5. Conclusions

To develop a model based on the stress state of a single grain, this study investigated the material removal mechanism and interference action in the grinding zone. The combined effect of the material removal and plastic stacking of a single grain has been considered.

Cutting experiment of single grain was carried out to obtain the critical cutting depth of the cutting and ploughing states. Results indicated that the variation trend of the grain-cutting efficiency (β) presents an S-shaped trend as the cutting depth (a_g) increases.

This study also developed and verified an improved grinding force predictive model. The average percentage of the deviation in the normal force is 4.19%, whereas that in the tangential force is 4.31%.

Acknowledgements

This research was financially supported by the following organizations: the National Natural Science Foundation of China (51975305 and 51905289), the Major Research Project of Shandong Province (2019GGX104040, 2019GSF108236, and 2018GGX103044), and the Shandong Provincial Natural Science Foundation of China (ZR2019PEE008), Major Science and technology innovation engineering projects of Shandong Province (2019JZZY020111), Applied basic research Youth Project of Qingdao science and technology plan (19-6-2-63-cg).

Conflict of interest

The authors declare no conflict of interest.

Author details

Yanbin Zhang[1,2], Changhe Li[1*], Yongjun Zhao[2*], Xin Cui[1], Xiufang Bai[3], Mingzheng Liu[1], Yali Hou[1], Min Yang[1], Naiqing Zhang[4], Heju Ji[5], Xiaoming Wang[1] and Teng Gao[1]

1 School of Mechanical Engineering, Qingdao University of Technology, Qingdao, China

2 MH Robot and Automation Co., Ltd., Weifang, China

3 School of Mechanical and Electrical Engineering, Qingdao Binhai University, Qingdao, China

4 Shanghai Jinzhao Energy Saving Technology Co. Ltd., Shanghai, China

5 Qingdao Dong Jia Textile Machinery Group Co., Ltd., Qingdao, China

*Address all correspondence to: sy_lichanghe@163.com; 18622115276@163.com

References

[1] Zhang YB, Li CH, Jia DZ, et al. Experimental evaluation of the lubrication performance of MoS_2/CNT nanofluid for minimal quantity lubrication in Ni-based alloy grinding. International Journal of Machine Tools & Manufacture. 2015;**99**:19-33. DOI: 10.1016/j.ijmachtools.2015.09.003

[2] Yang M, Li CH, Zhang YB, et al. Research on microscale skull grinding temperature field under different cooling conditions. Applied Thermal Engineering. 2017;**126**:525-537. DOI: 10.1016/j.applthermaleng.2017.07.183

[3] Jia DZ, Li CH, Zhang YB, et al. Experimental evaluation of surface topographies of NMQL grinding ZrO_2 ceramics combining multiangle ultrasonic vibration. The International Journal of Advanced Manufacturing Technology. 2019;**100**(1-4):457-473. DOI: 10.1007/s00170-018-2718-y

[4] Wu WT, Li CH, Yang M, et al. Specific Energy and G ratio of Grinding Cemented Carbide under Different Cooling and Lubrication Conditions. The International Journal of Advanced Manufacturing Technology. 2019;**105**(1-4):67-82. DOI: 10.1007/s00170-019-04156-5

[5] Wang YG, Li CG, Zhang YB, et al. Comparative evaluation of the lubricating properties of vegetable-oil-based nanofluids between frictional test and grinding experiment. Journal of Manufacturing Processes. 2017;**26**:94-104. DOI: 10.1016/j.jmapro.2017.02.001

[6] Li HN, Xie KG, Wu B, Zhu WQ. Generation of textured diamond abrasive tools by continuous-wave CO_2 laser: Laser parameter effects and optimisation. Journal of Materials Processing Technology. 2020;**275**:116279. DOI: 10.1016/j.jmatprotec.2019.116279

[7] Li HN, Yang Y, Zhao YJ, et al. On the periodicity of fixed-abrasive planetary lapping based on a generic model. Journal of Manufacturing Processes. 2019;**44**:271-287. DOI: 10.1016/j.jmapro.2019.05.036

[8] Gao T, Zhang XP, Li CH, et al. Surface morphology evaluation of multi-angle 2D ultrasonic vibration integrated with nanofluid minimum quantity lubrication grinding. Journal of Manufacturing Processes. 2020;**51**:44-61. DOI: 10.1016/j.jmapro.2020.01.024

[9] Zhang YB, Li CH, Jia DZ, et al. Experimental evaluation of MoS2 nanoparticles in jet MQL grinding with different types of vegetable oil as base oil. Journal of Cleaner Production. 2015;**87**:930-940. DOI: 10.1016/j.jclepro.2014.10.027

[10] Yang M, Li CH, Zhang YB, et al. Predictive model for minimum chip thickness and size effect in single diamond grain grinding of zirconia ceramics under different lubricating conditions. Ceramics International. 2019;**45**(12):14908-14920. DOI: 10.1016/j.ceramint.2019.04.226

[11] Yang M, Li CH, Zhang YB, et al. Maximum undeformed equivalent chip thickness for ductile-brittle transition of zirconia ceramics under different lubrication conditions. International Journal of Machine Tools and Manufacture. 2017;**122**:55-65. DOI: 10.1016/j.ijmachtools.2017.06.003

[12] Werner G. Influence of work material on grinding forces. CIRP Annals. 1978;**27**:243-248

[13] Malkin S, Hwang TW. Grinding mechanisms for ceramics. CIRP Annals. 1996;**45**(2):569-580. DOI: 10.1016/s0007-8506(07)60511-3

[14] Li L, Fu J, Peklenik J. A study of grinding force mathematical model.

CIRP Annals. 1980;**29**(1):245-249. DOI: 10.1016/S0007-8506(07)61330-4

[15] Younis M, Sadek MM, El-Wardani T. A new approach to development of a grinding force model. Journal of Engineering Industry. 1987;**109**(4): 306-313

[16] Tang J, Du J, Chen Y. Modeling and experimental study of grinding forces in surface grinding. Journal of Materials Processing Technology. 2009;**209**(6): 2847-2854. DOI: 10.1016/j.jmatprotec.2008.06.036

[17] Hecker RL, Liang SY, Wu XJ, et al. Grinding force and power modeling based on chip thickness analysis. The International Journal of Advanced Manufacturing Technology. 2007;**33** (5-6):449-459. DOI: 10.1007/s00170-006-0473-y

[18] Lang X, He Y, Tang J, et al. Grinding force model based on prominent height of abrasive submitted to Rayleigh distribution. Journal of Central South University (Science and Technology). 2014;**45**(10):3386-3391

[19] Zhang J, Ge P, Zhang L. Research on the grinding force based on the probability statistics. China Mechanical Engineering. 2007;**18**(20): 2399-2402

[20] Chang CH, Wang J-JJ. A stochastic grinding force model considering random grit distribution. International Journal of Machine Tools and Manufacture. 2008;**48**(12-13): 1335-1344. DOI: 10.1016/j.ijmachtools.2008.05.012

[21] Cheng J, Wu J, Gong YD, et al. Grinding forces in micro slot-grinding (MSG) of single crystal sapphire. International Journal of Machine Tools and Manufacture. 2017;**112**:7-20. DOI: 10.1016/j.ijmachtools.2016.10.004

[22] Zhou M, Zheng W. A model for grinding forces prediction in ultrasonic vibration assisted grinding of SiCp/Al composites. The International Journal of Advanced Manufacturing Technology. 2016;**87**(9-12):3211-3224. DOI: 10.1007/s00170-016-8726-x

[23] Jiang Z, Yin Y, Wang Q, et al. Predictive modelling of grinding force considering wheel deformation for toric fewer-axis grinding of large complex optical mirrors. Journal of Manufacturing Science and Engineering. 2016;**138**(6):061008. DOI: 10.1115/1.4032084

[24] Li HN, Yu TB, Wang ZX, et al. Detailed modeling of cutting forces in grinding process considering variable stages of grain-workpiece micro interactions. International Journal of Mechanical Sciences. 2017;**126**:319-339. DOI: 10.1016/j.ijmecsci.2016.11.016

[25] Younis MA, Alawi H. Probabilistic analysis of the surface grinding process. Transactions of the Canadian Society for Mechanical Engineering. 1984;**8**(4): 208-213. DOI: 10.1139/tcsme-1984-0031

[26] Zhang Y, Li C, Yang M, et al. Experimental evaluation of cooling performance by friction coefficient and specific friction energy in nanofluid minimum quantity lubrication grinding with different types of vegetable oil. Journal of Cleaner Production. 2016;**139**: 685-705. DOI: 10.1016/j.jclepro.2016.08.073

[27] Cui X, Li C, Zhang Y, Jia D, Zhao Y, Li R, et al. Tribological properties under the grinding wheel and workpiece interface by using graphene nanofluid lubricant. The International Journal of Advanced Manufacturing Technology. 2019;**104**(9-12):3943-3958. DOI: 10.1007/s00170-019-04129-8

[28] Zhang ZY, Wang B, Kang RK, et al. Changes in surface layer of silicon wafers from diamond scratching. CIRP Annals. 2015;**64**(1):349-352. DOI: 10.1016/j.cirp.2015.04.005

[29] Cao JG, Wu YB, Li JY, et al. A grinding force model for ultrasonic assisted internal grinding (UAIG) of SiC ceramics. The International Journal of Advanced Manufacturing Technology. 2015;**81**(5-8):875-885. DOI: 10.1007/s00170-015-7282-0

[30] Zhang Z, Guo D, Wang B, et al. A novel approach of high speed scratching on silicon wafers at nanoscale depths of cut. Scientific Reports. 2015;**5**:16395. DOI: 10.1038/srep16395

[31] Zhang Z, Guo L, Cui J, et al. Nanoscale solely amorphous layer in silicon wafers induced by a newly developed diamond wheel. Scientific Reports. 2016;**6**:35269. DOI: 10.1038/srep35269

[32] Zhang J, Song Y, Huo F, et al. Nanoscale material removal mechanism of soft-brittle HgCdTe single crystals under nanogrinding by ultrafine diamond grits. Tribology Letters. 2012;**46**(1):95-100. DOI: 10.1007/s11249-012-9924-9

[33] Ding W, Dai J, Zhang LC, et al. An investigation on the chip formation and forces in the grinding of Inconel718 alloy using the single-grain method. Issues in Mental Health Nursing. 2014; **28**:1-2. DOI: 10.1016/j.ijmachtools.2017.06.002

[34] Jin T, Stephenson DJ. Heat flux distributions and convective heat transfer in deep grinding. International Journal of Machine Tools and Manufacture. 2006;**46**(14):1862-1868. DOI: 10.1016/j.ijmachtools.2005.11.004

[35] Zhang J, Wang Y, Wang Y. Experiment and constitutive model of rate-dependent behavior of titanium alloy TC11. The Chinese Journal of Nonferrous Metals. 2017;**27**(07): 1369-1375. DOI: 10.19476/j.ysxb.1004.0609.2017.07.07

[36] Xun C, Rowe WB. Analysis and simulation of the grinding process. Part II: Mechanics of grinding. International Journal of Machine Tools and Manufacture. 1996;**36**(8):883-896. DOI: 10.1016/0890-6955(96)00117-4

[37] Du HJ, Rui YN, Wang R, et al. MATLAB-based simulation method of surface's three-dimensional model in grinding process. Modern Manufacturing Engineering. 2009;(02): 48-51. DOI: CNKI:SUN:XXGY.0.2009-02-013

[38] Öpöz TT, Chen X. Effect of different parameters on grinding efficiency and its monitoring by acoustic emission. Production & Manufacturing Research. 2016;**4**(1): 190-208. DOI: 10.1080/21693277.2016.1255159

[39] Lindsay RP, Hahn RS. Principle of Grinding, Part I, II. Machinery. 1971

[40] Zhang Y, Li C, Ji H, et al. Analysis of grinding mechanics and improved predictive force model based on material-removal and plastic-stacking mechanisms. International Journal of Machine Tools and Manufacture. 2017; **122**:81-97. DOI: 10.1016/j.ijmachtools.2017.06.002

[41] Agarwal S, Venkateswara Rao P. Predictive modeling of force and power based on a new analytical undeformed chip thickness model in ceramic grinding. International Journal of Machine Tools and Manufacture. 2013; **65**:68-78. DOI: 10.1016/j.ijmachtools.2012.10.006

[42] Cai R, Qi HS, Cai GQ. Active cutting edges in vitrified CBN grinding wheels. Key Engineering Materials. 2006;**304**: 1-7. DOI: 10.4028/www.scientific.net/KEM.304-305.1

Chapter 7

Kerosene-Water Multiphase Flow in Vertical and Inclined Pipes

Faik Hamad, Nadeem Ahmed Sheikh and Muzaffar Ali

Abstract

This chapter presents the volume fraction distribution of kerosene-water two-phase flow in vertical and inclined pipes. The study of liquid-liquid two-phase flow is very significant to oil industry and many other processes in industry where two liquids are mixed and flow together. Pitot tube and optical probes are used for the measurement of velocity of water and volume fraction. The experimental measurements of the local parameters demonstrate that the single-phase and two-phase flows reached the fully developed axisymmetric conditions at $L/D \geq 48$ (L, pipe length; D, pipe diameter). The results also showed the severe asymmetry distributions of the volume fraction at the entrance region ($L/D = 1$) downstream the bend and in the inclined pipe. The comparison of volume fraction profiles with void fraction profiles indicated a significant difference in their shapes. The results also showed that the kerosene accumulated at the upper wall of the inclined pipe and the distribution improved by increasing the volumetric quality.

Keywords: volume fraction, kerosene-water two-phase flow, vertical and inclined pipes, optical probe, pitot tube

1. Introduction

Multiphase flows are important for the design of steam/water flow in steam generators, jet engines, condensers, extraction and distillation processes, gas and oil mixture in pipelines, and refrigeration systems. The mixture of two immiscible liquids is characterized by the existence of interfaces between the two fluids, associated with a discontinuity of properties across the interface. The single-phase flow is traditionally classified into laminar, transitional, and turbulent flows according to the flow Reynolds number. The two-phase flow in vertical pipe can be classified, according to the geometry of the interfaces.

The primary condition for all two-phase flows is specified by the volumetric quality β, which is defined as:

$$\beta = \frac{Q_d}{Q_c + Q_d} \tag{1}$$

where Q_d is the flow rate of the dispersed phase and Q_c is the flow rate of the continuous phase.

For a pipe of radius R, the corresponding (area averaged) superficial velocities are defined as:

IntechOpen

Continuous phase superficial velocity:

$$\overline{U}_{cs} = \frac{Q_c}{\pi R^2} \tag{2}$$

Dispersed phase superficial velocity:

$$\overline{U}_{ds} = \frac{Q_d}{\pi R^2} \tag{3}$$

For both gas-liquid and liquid-liquid flow systems, the continuous phase is usually water.

In spite of the large number of published work in multiphase flow area, the publication on using local probe measurements for liquid-liquid flow is very limited compared to the gas-liquid two-phase flow. The purpose of this chapter is to publish some data on volume fraction profiles for liquid-liquid flow in vertical and inclined pipes. The following data are presented in this chapter: (i) the void fraction distribution for gas-liquid two-phase flow in vertical pipe, (ii) the volume fraction distribution for flow development of kerosene-water flow in vertical pipe, and (iii) the volume fraction distribution for the fully developed kerosene-water flow in vertical and inclined pipes.

2. Void fraction/volume fraction definition

Most experimental results for the void fraction α (volume fraction for liquid-liquid flow) have been obtained by a point sensor, which was used to discriminate in time between the two phases. Experimentally the void fraction α has been evaluated from the time record from such a probe as:

$$\alpha = \frac{\Sigma \Delta t_d}{T} \tag{4}$$

where $\Sigma \Delta t_d$ is the time the probe is located in the dispersed phase and T is the total sampling time used for record. It should be noticed that in many investigations, reference is made to the (average) void fraction $\overline{\alpha}$. The proper reference would have been the volumetric quality (β) obtained for gas-liquid systems by the quick closing valve method, X-ray or neutron techniques. For liquid-liquid systems, β can be obtained by measuring the two flow rates Q_d and Q_c.

Experimental studies of the phase distributions in concurrent two-phase upflow in vertical pipes present a complex picture that has not yet been systematically evaluated. The common flow patterns for vertical upward flow, in which both phases flow upwards in a circular tube, are shown in **Figure 1**. As the volume flow rate of gas increased for constant water flow rate, the flow patterns would vary. The following types of flow patterns can be found in vertical pipes:

i. Bubbly flow: bubbles of gas or liquid in a continuous liquid phase appear, and the size of the bubbles can be very small or large.

ii. Slug flow: in this type a bullet-shaped plug of gas is formed from many bubbles concentrated in one part to make larger bubbles, which approach the diameter of the pipe. The liquid phase is in continuous flow.

iii. Churn flow: braking down of large vapor bubbles in plug flows form the churn flow. This is a highly oscillatory flow, and there is tendency for each phase to be continuous with irregular interfaces.

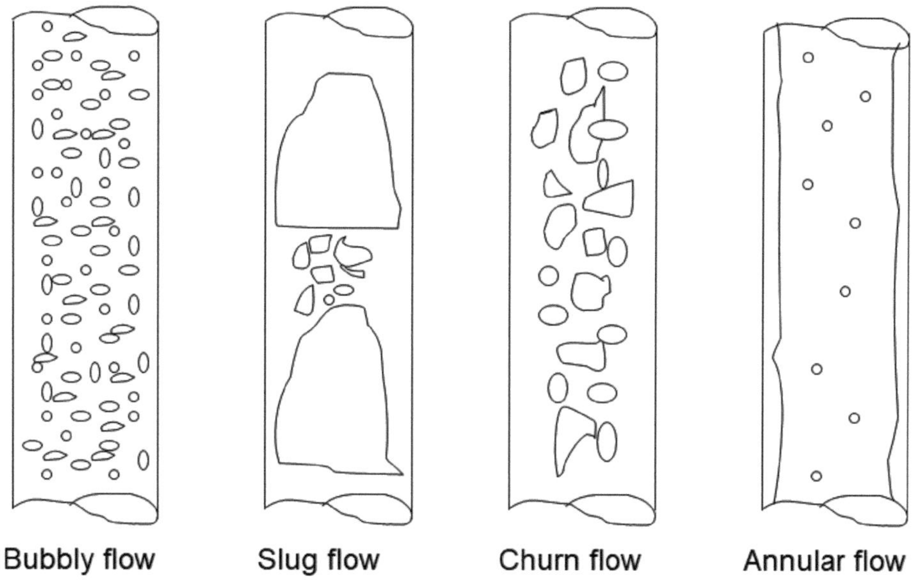

Figure 1.
Flow patterns in vertical upward flow in a pipe [1].

iv. Annular flow: the liquid forms a film around the wall of the tube. The gas phase flows in the centre.

3. Gas: liquid void fraction distribution in vertical pipes

A significant number of measurements have been made for upflow in vertical pipes. Several investigators, e.g. Malnes [2], Serizawa et al. [3], Michiyoshi and Serizawa [4], Wang et al. [5] and Liu and Bankoff [6], have observed the peaking phenomenon of the local void fraction near the wall as shown in **Figure 2**. Some investigators, such as Van der Welle [7], Moujaes and Dougall [8] and Johnson and White [9], have observed a maximum void fraction at the centreline as shown in **Figure 3**. Other researchers, such as Nakoryakov et al. [10], Spindler et al. [11] and

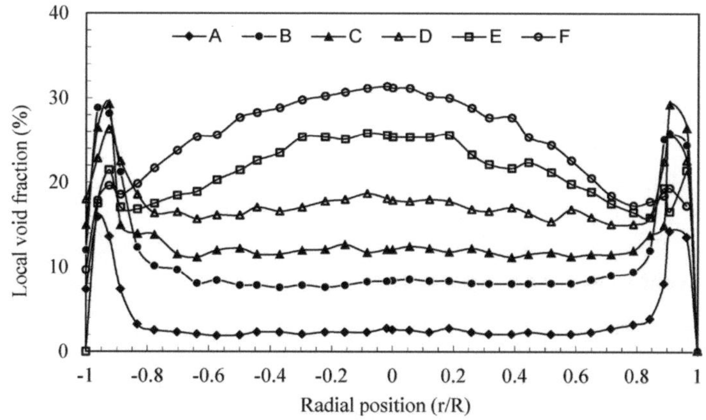

Figure 2.
Void fraction distributions for air-water upflow in a pipe with a continuous phase mean velocity = 1.03 m/s, Z/D = 30. A–D are bubbly flow, and E and F are slug flow [3].

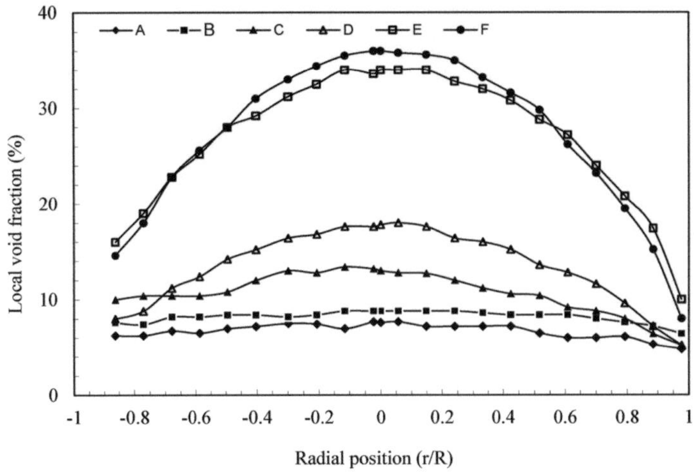

Figure 3.
Void fraction distributions for air-water upflow in a pipe with different flow rates, Z/D = 42.5. A–D are bubbly flow, and E and F are transitional flow [9].

Liu [12], observed both wall and centreline peaking void fraction distributions for two-phase flow. The actual void fraction distribution configurations have been found to depend on the initial conditions: bubble size and flow rates, physical properties of the fluids, and the test section condition geometry.

4. Liquid: liquid volume fraction

Compared to the large number of publications on gas-liquid flows, less work have been published on liquid-liquid flows.

Most of the papers on liquid-liquid mixture flow were published by research group at the University of Bradford ([13, 14]; Hamad et al. [15]; Hamad and Bruun [16]). Most of these papers focused on the development of optical techniques for kerosene-water upward flow in vertical pipes. However, Farrar and Bruun [13] highlighted the problem of the severe asymmetry, and the swirl generated upstream the inlet due the existence of 90° bend as part of the experimental facility.

Zhao et al. [17] used a double-sensor conductivity probe to measure the local oil phase fraction distribution for flow in a vertical pipe at L/D = 72. They found that the volume fraction profiles were uniform for $\beta < 9.2\%$ and changed into wall peak for $\beta > 9.2\%$. The local oil phase fraction profiles at (a) constant water flow rate ($J_w = 0.33$ m/s) and (b) constant oil flow rate ($J_o = 0.066$ m/s) are given in **Figure 4**.

A comprehensive experimental data on kerosene-water two-phase flow were published by Hamad et al. [18, 19] and Hamad et al. [20] in vertical and inclined pipes. A summary of the results from each paper is given in the following sections.

4.1 Development of kerosene-water flow in vertical pipe

Hamad et al. [18] studied the flow development in a vertical pipe of 77.8 mm inner diameter and 4500 mm length downstream of a 90° bend experimentally at L/D = 1, 16, 38 and 54 using the experimental facility in **Figure 5**. Single-phase (water) flow measurements were made to check the establishment of fully developed symmetrical flow conditions. **Figure 6** shows the radial distribution of axial velocity in the plane parallel to the bend at different L/D ratios. Two values of $\overline{U}_{ws}$

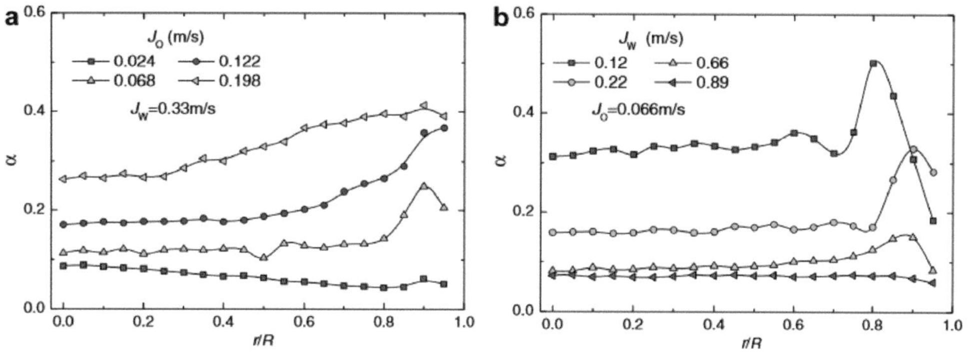

Figure 4.
Oil volume fraction profiles at (a) constant water flow rate (J_w = 0.33 m/s) and (b) constant oil flow rate (J_o = 0.066 m/s) [17].

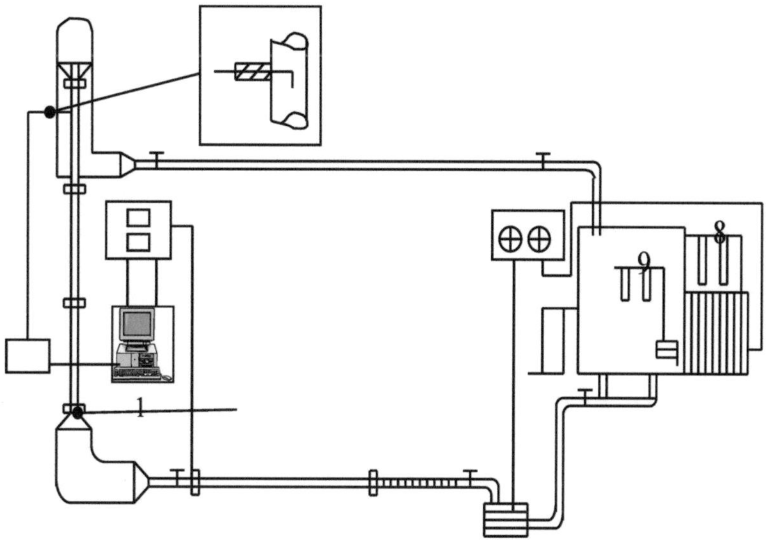

Figure 5.
Schematic diagram of two-phase flow experimental facility [18].

are used (0.44 m/s (*Re* = 33,800) and 0.77 m/s (*Re* = 60,000)). The results show that water velocity distributions become fully developed at $L/D > 48$. The empirical power law velocity distribution given in Eq. (5) for single-phase turbulent flow [21, 22] was also included in **Figure 6** to confirm the accuracy of the measurements:

$$\overline{U} = \overline{U}_{cl}(1 - r/R)^{1/n} \tag{5}$$

Then, the kerosene was introduced to perform volume fraction measurements using optical probe [14] at four different axial positions at L/D = 1, 16, 38 and 54 downstream of the pipe bend. Three different flow conditions are considered: Case 1: water superficial velocity, $\overline{U}_{ws}$ = 0.44 m/s, and volumetric quality, β = 9.2%; Case 2: $\overline{U}_{ws}$ = 0.44 m/s and β = 18.6% and one high $\overline{U}_{ws}$ condition; and Case 3: $\overline{U}_{ws}$ = 0.77 m/s and β = 18.6%. For Case 1, the axisymmetric distribution is very poor at L/D = 1 (**Figure 7(a)**) with high volume fraction values of 20% near the inner side of the bend to the lower value near the outer wall of the bend of 4%. For Case 2, increasing β to 18.6% for the same $\overline{U}_{ws}$ (**Figure 7(a)**) improves the axisymmetric distribution across the pipe. For Case 3, the axisymmetric distribution improved

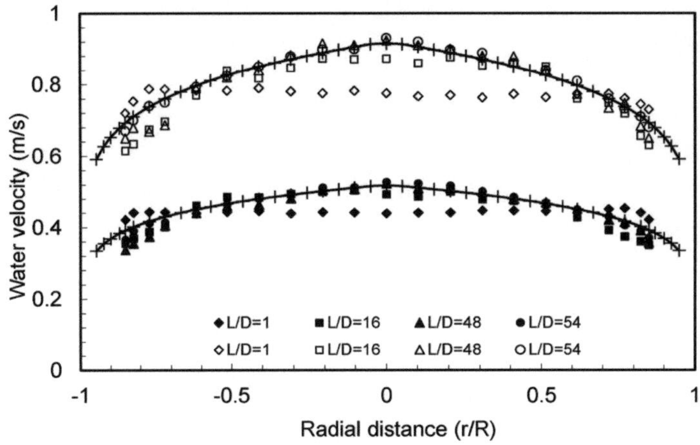

Figure 6.
Local single-phase velocity distribution at different L/D ratios for average water velocity = 0.44 m/s (solid symbols) and 0.77 m/s (open symbols). The power law velocity distribution (solid line) is also included [18].

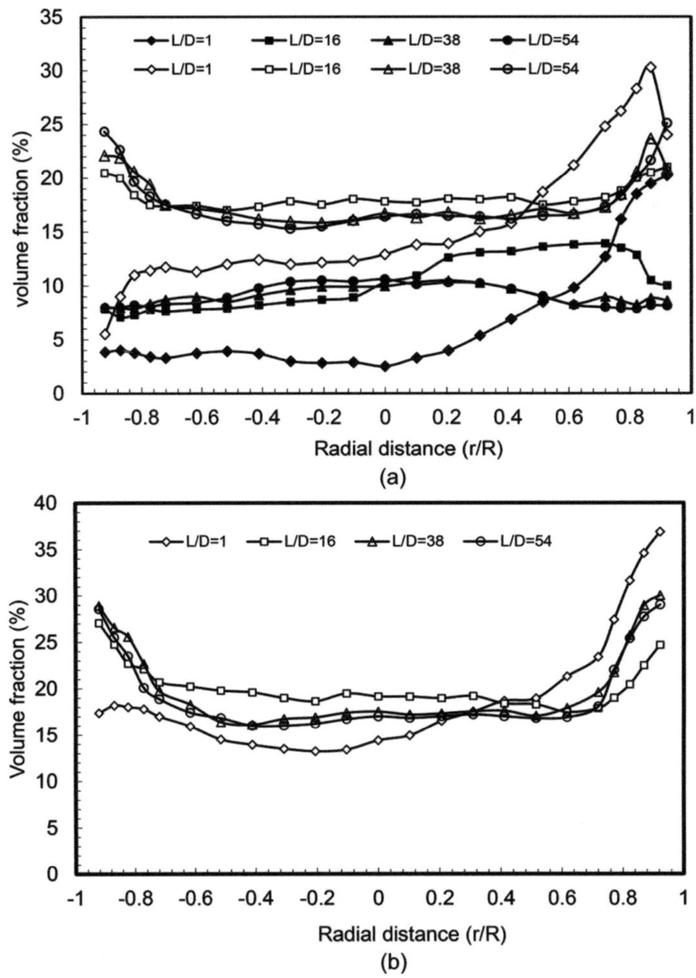

Figure 7.
Volume fraction distributions at different (a) L/D for $\overline{U}_{ws}$ = 0.44 m/s and β = 9.2% (solid symbols) and 18.6% (open symbols) [18], (b) L/D *for $\overline{U}_{ws}$ = 0.77 m/s and β = 18.6% [18].*

further by increasing $\bar{U}_{ws}$ for the same β (**Figure 7(b)**). The main conclusion is that the axisymmetric becomes better for higher β and $\bar{U}_{ws}$. The change in volume fraction distribution may be attributed to the improvement of the mixing process of the kerosene with water which acts against the effect of the buoyancy force and the centrifugal force at the outlet of the bend. However, the distribution becomes nearly symmetrical at L/D = 16. There appear to be no significant differences between the distributions at L/D = 38 and 54, which suggests that fully developed, symmetrical condition was achieved.

4.2 Fully developed flow of kerosene-water flow in vertical pipe

Hamad et al. [19] studied the flow of kerosene-water upward flow in a vertical pipe at (L/D = 54) using optical probes. The effects of $\bar{U}_{ws}$ and β on radial volume fraction distribution [$\alpha(r)$] of two-phase flow parameters were investigated.

The local volume fraction is calculated from the output of the leading sensor of the dual optical probe by determining the average drop residence time using the procedure described in Hamad et al. [14]. Comprehensive measurements

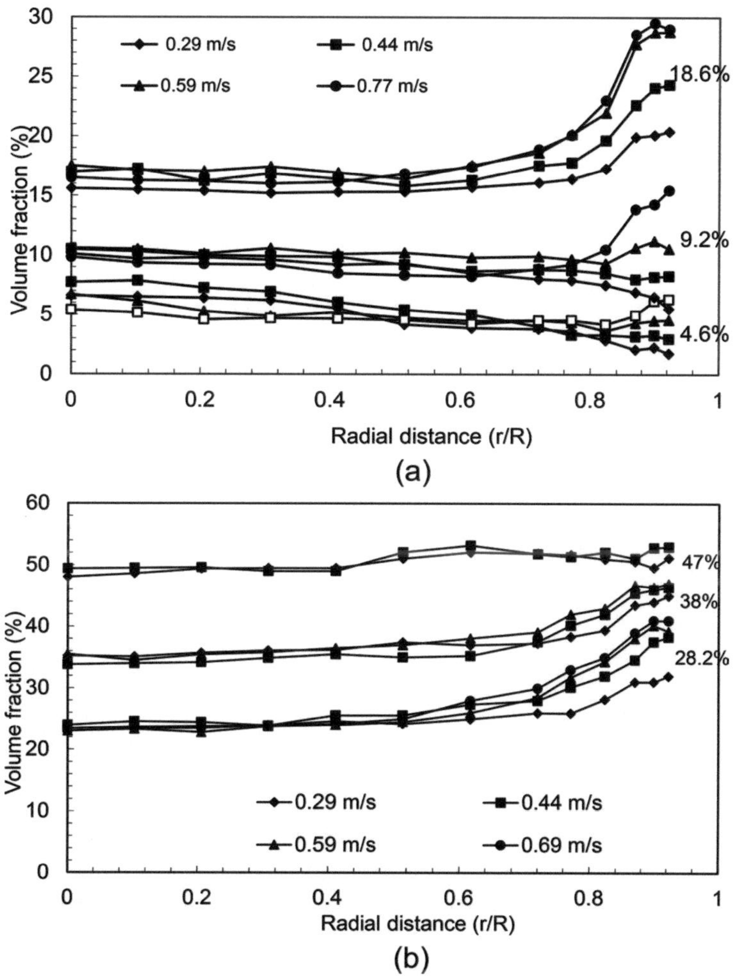

Figure 8.
Volume fraction profiles for different (a) $\bar{U}_{ws}$ and β = 4.6, 9.2 and 18.6% [19], (b) $\bar{U}_{ws}$ and β = 28.2, 38 and 47% [19].

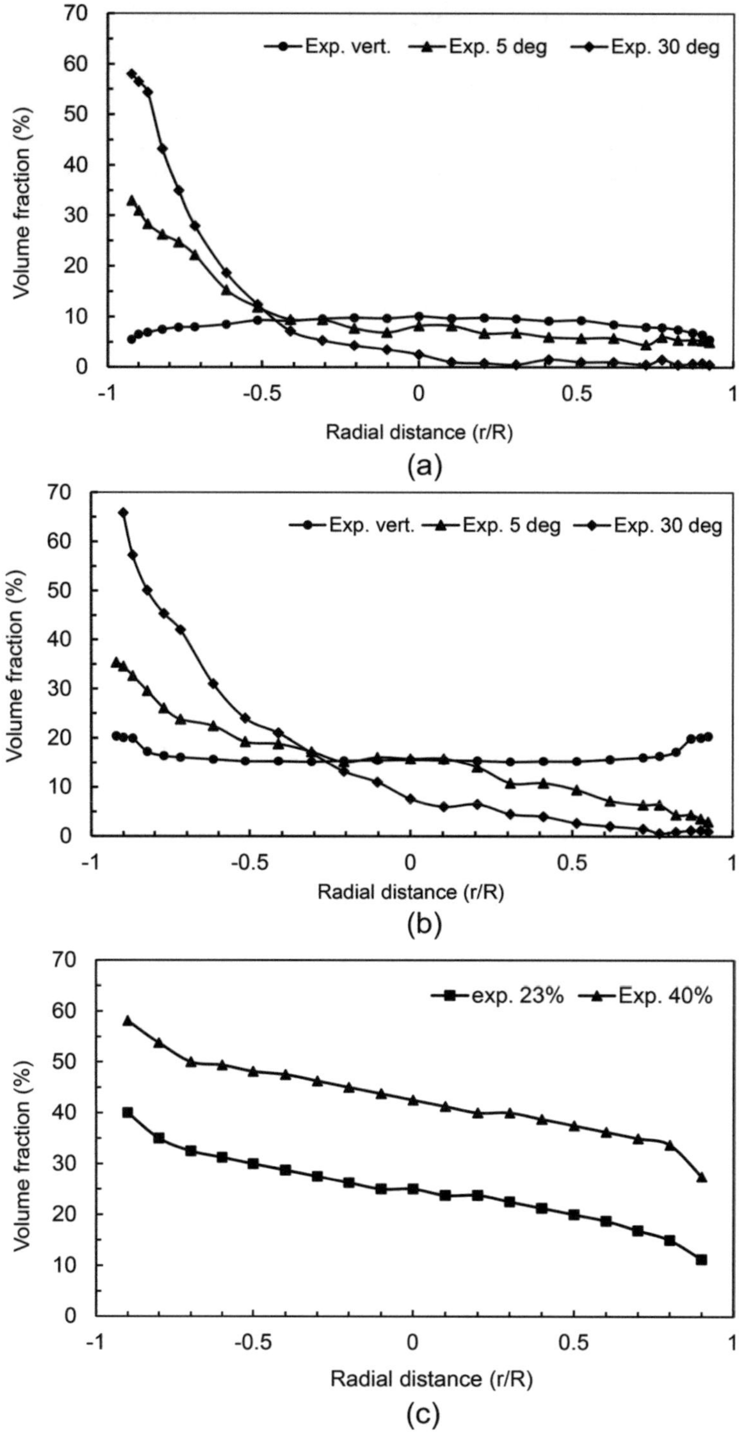

Figure 9.
The effect of pipe inclination on volume fraction distribution (a) (U_{sw} = 0.29 m/s, β = 9.2%) [20], (b) (U_{sw} = 0.29 m/s, β = 18.6%) [20]. (c) the α(r) distribution from Vigneaux et al. [24].

of $\alpha(r)$ were performed for a number of β values in the range of 4.6–47% and constant $\overline{U}_{ws}$ of 0.29, 0.44, 0.59, 0.69 and 0.77 m/s. The $\alpha(r)$ profiles have been plotted together for various values of $\overline{U}_{ws}$ for each value of β as shown in **Figure 8(a)** and **(b)**.

As the $\alpha(r)$ profile primarily reflects the kerosene content in the mixture flow, it follows that the related $\alpha(r)$ profile sets for β = 4.6%, 9.2%, 18.6%, 28.2%, 38% and 47% are centred around these values. The graphs also show distinct variations, both within each β group and between groups with different β values.

The results from **Figure 8(a)** and **(b)** show that increasing $\overline{U}_{ws}$ with low β (<20%) will change the $\alpha(r)$ profiles from convex shape with peak at the pipe centreline to flat shape and then to concave shape with peak near the wall. For moderate β (20–40%), the $\alpha(r)$ profiles have a concave shape for different $\overline{U}_{ws}$ with peak near the wall which has high values for higher $\overline{U}_{ws}$. In the case of $\beta \approx 50\%$, the $\alpha(r)$ profile shapes are flat for the two cases in **Figure 8(b)**.

The $\alpha(r)$ profiles from centreline which peaked to uniform to wall peaked and then to uniform can be attributed to the change in lift force due to the change in drop diameter, slip velocity and radial velocity distribution of both phases. The present finding is supported by the results for liquid-liquid flows from Zhao et al. [17] and Hua et al. [23] for the same range of $\overline{U}_{ws}$ and β.

4.3 Kerosene-water flow in inclined pipe

Hamad et al. [20] used an optical probe to study the kerosene-water flow inclined at 5° and 30° from vertical at L/D = 54. The volume fraction was measured for $\overline{U}_{ws}$ = 0.29 m/s and β = 9.2% and 18.6%.

Figure 9(a) shows the radial $\alpha(r)$ distributions of the volume fraction, $\alpha(r)$ for 0°, 5° and 30° inclination angles at $\overline{U}_{ws}$ of 0.29 m/s and two values of β = 9.2 and 18.6%.

The results in **Figure 9(a)** and **(b)** show that the inclination has a significant influence on the distribution of $\alpha(r)$. The kerosene drops were separated from the water accumulated at the upper zone of the pipe due to the gravity effect. The effect of increasing β in an inclined pipe leads to dispersion of the drops to the lower zone of the pipe due to the recirculation cells of the moving droplet swarms.

The present results are supported by the findings reported by Vigneaux et al. [24] and Flores et al. [25]. **Figure 9(c)** presents the two sets of experimental data reported by Vigneaux et al. [24] in a pipe inclined at 15° from vertical. In the first case, β = 23%, and U_{SW} = 0.27 m/s, and in the second case, β = 40%, and U_{SW} = 0.21 m/s.

5. Conclusion

The results on void fraction profiles from literature show the complexity of the flow behaviour. It is reflected in different types of profiles due to the local interaction between the bubbles and the continuous phase. This may be attributed to the various forces at interface between the phases including drag, lift and virtual force as well as the size of bubbles and compressibility effect. In contrast, the volume fraction profiles for liquid-liquid two-phase flow have similar shapes. This behaviour may be attributed to smaller drops, smaller density ratio, smaller slip velocity and the incompressible nature of the liquids.

The results show that fully developed condition for liquid-liquid flow can be achieved at lower L/D compared to gas-liquid flow. This is due to incompressible nature of liquid drops which have the same volume compared to the gas bubbles which expand continuously due to the pressure drop in flow direction.

Acknowledgements

Authors are thankful to their parent institutions for providing support for the research.

Author details

Faik Hamad[1*], Nadeem Ahmed Sheikh[2] and Muzaffar Ali[3]

1 School of Science, Engineering, and Design, Teesside University, UK

2 Department of Mechanical Engineering, Faculty of Engineering and Technology, International Islamic University, Islamabad, Pakistan

3 Department of Energy Engineering, University of Engineering and Technology Taxila, Pakistan

*Address all correspondence to: f.hamad@tees.ac.uk

References

[1] Hamad FA. A study of the continuous and dispersed phase parameters in kerosene-water two-phase up-flow in vertical pipe [PhD Thesis]. UK: University of Bradford; 2001

[2] Malnes D. Slip Ratios and Friction Factors in Bubble Flow Regime in Vertical Pipes, Report No. KR-100. Kjeller, Norway: Institute for Atomenergi; 1966

[3] Serizawa A, Kataoka I, Michiyoshi I. Turbulence structure of water bubbly flow: II. Local properties. International Journal of Multiphase Flow. 1975;**2**:235-246

[4] Michiyoshi I, Serizawa A. Turbulence in two-phase bubbly flow. Nuclear Engineering and Design. 1986;**95**:235-267

[5] Wang SK, Lee SJ, Jones OC, Lahey RT. 3-D turbulence structure and phase distribution measurements in bubbly two-phase flow. International Journal of Multiphase Flow. 1987;**13**:327-343

[6] Liu TJ, Bankoff SG. Structure of air-water bubbly flow in vertical pipe-11. Void fraction, bubble velocity, and bubble size distribution. International Journal of Heat and Mass Transfer. 1993;**36**:1061-1072

[7] Van der Welle. Void fraction, bubble velocity, and bubble size in two-phase flow. International Journal of Multiphase Flow. 1985;**11**:317-345

[8] Moujaes S, Dougall RS. Experimental investigation of cocurrent two-phase flow in a vertical rectangular channel. Canadian Journal of Chemical Engineering. 1987;**65**:705-715

[9] Johnson AB, White DB. Experimental determination of gas migration velocities with non-Newtonian fluids. International Journal of Multiphase Flow. 1993;**19**:921-941

[10] Nakoryakov VE, Kashinsky ON, Burdukov AP, dnoral VP. Local characteristics of upward gas-liquid flows. International Journal of Multiphase Flow. 1981;**7**:63-81

[11] Spindler K, Bierer M, Lorenz G, Erhard A, Hahne E. Measurements in vertical gas-liquid two-phase flows using an optical fiber probe. In: Proceedings of the First Word Conference on Experimental Heat Transfer, Fluid Mechanics and Thermodynamics. Dubrovnik, Yugoslavia; 1988. pp. 348-357

[12] Liu TJ. Bubble size and entrance length effects on void development in a vertical channel. International Journal of Multiphase Flow. 1993;**19**:99-113

[13] Farrar B, Bruun HH. A computer based hot-film technique used for flow measurements in a vertical kerosene-water pipe flow. International Journal of Multiphase Flow. 1996;**22**:733-752

[14] Hamad FA, Imberton F, Bruun HH. An optical probe for measurements in liquid-liquid two-phase flow. Measurement Science and Technology. 1997;**8**:1122-1132

[15] Hamad FA, Bruun HH. Evaluation of bubble/drop velocity by a single normal hot-film placed in a two-phase flow. Measurement Science and Technology. 2000;**11**:11-19

[16] Hamad FA, Pierscionek BK, Bruun HH. A dual optical probe for volume fraction, drop velocity and drop size measurements in liquid-liquid two-phase flow. Measurement Science and Technology. 2000;**11**:1307-1318

[17] Zhao D, Guo L, Hu X, Zhang X, Wang X. Experimental study on local characteristics of oil-water dispersed

flow in a vertical pipe. International Journal of Multiphase Flow. 2006;**32**:1254-1268

[18] Hamad FA, He S, Khan MK, Bruun HH. Development of kerosene–water two-phase up-flow in a vertical pipe downstream of a 90° bend. The Canadian Journal of Chemical Engineering. 2013a;**91**(2):354-367

[19] Hamad FA, Khan MK, Bruun HH. Experimental study of kerosene–water two-phase flow in a vertical pipe using hot-film and dual-optical probe bend. The Canadian Journal of Chemical Engineering. 2013b;**91**(7):1296-1311

[20] Hamad FA, Albarzenji D, Ganesan p. Study of kerosene-water two-phase flow characteristics in vertical and inclined pipes. The Canadian Journal of Chemical Engineering. 2014;**92**(5):905-917

[21] Munson BR, Young DF, Okiishi TH. Fundamentals of Fluid Mechanics. 4th ed. New York, USA: John Willey & Sons, Inc; 2002

[22] Schlichting H. Boundary Layer Theory. New York, USA: McGraw-Hill; 1979

[23] Hua L, Mi W, Ying-Xiang W, Yi-Xin M, Richard W. Measurements of oil volume fraction and velocity distributions in vertical oil-in-water flows using ERT and a local probe. Journal of Zhejiang University. Science. 2005;**6A**(12):1412-1415

[24] Vigneaux P, Chenais P, Hulin JP. Liquid-liquid flows in an inclined pipe. AICHE Journal. 1988;**34**:781-789

[25] Flores J, Chen XT, Brill JP. Characteristics of oil-water flow patterns in vertical and deviated well. SPE Production & Facilities. 1999;**14**(2):94-101

Chapter 8

Heat Transfer and Fluid Flow Investigations in PDMS Microchannel Heat Sinks Fabricated by Means of a Low-Cost 3D Printer

Inês Maia, Cesar Rocha, Pedro Pontes, Vanessa Cardoso, João M. Miranda, Ana S. Moita, G. Minas, António L.N. Moreira and Rui Lima

Abstract

Polydimethylsiloxane (PDMS), due to its remarkable properties such as optical transparency and ability to easily mold, is one of the most popular polymers used in micro- and nanofluidics. Furthermore, 3D printing technology due to its low cost and simplicity is also gaining a great interest among the microfluidic community. In this work, the potential of 3D printing is shown to produce microfluidic devices, their ability for studying flows and heat transfer of nanofluids, and their applicability as a heat sink device. The low-cost fused deposition modeling 3D printing technique was combined with a PDMS casting technique for the microfluidic device fabrication. The potential of this technique was experimentally demonstrated by fluid flow and heat transfer investigations using different fluids, such as distilled water-, alumina (Al_2O_3)-, and iron oxide (Fe_3O_4)-based nanofluids. The simplicity, low-cost, and unique features of the proposed heat sink device may provide a promising way to investigate nanofluids' flow and heat transfer phenomena that are not possible to be studied by the current traditional systems.

Keywords: microfluidics, 3D printing, microchannel heat sinks, nanofluids, heat transfer, electronics cooling

1. Introduction

The continuous investigation on strategies for size reduction while maintaining, or even increasing, power of technological devices demands cooling systems with higher thermal efficiency and smaller sizes. One approach relied on the modification of heat sinks by incorporating microchannels to increase the heat exchange surface. Despite being a well-adopted strategy, its complex configurations proved to be difficult to manufacture [1, 2]. Other strategies focused on the type of the used fluids for the cooling process. Dielectric fluids [1, 3], two-phase fluids [3, 4], and nanofluids (NFs) [5–8] have been thoroughly investigated. Among these three, the nanofluids, which are fluids comprised of particles, with size ranging from 1 to 100 nm, suspended in a base fluid, have been reported as presenting a better thermal conductivity than the base fluid [8, 9]. Nevertheless, some challenges regarding their usage, such as agglomeration, long-term stability, and high-costs, still need further investigation [8, 10–12]. The most commonly used nanoparticles (NPs) for NFs are metallic, such as Cu, Ag, Au, and Fe, or non-metallic such as Al_2O_3, CuO, TiO_2 SiC, and carbon nanotubes. Many authors reported that the heat transfer of NFs is influenced by several factors such as shape, dimensions, volume fractions in the suspensions, and the thermal properties of the particle materials [11, 13, 14]. Therefore, enhancement in heat transfer was also reported, but only for small concentrations of NPs [8, 9].

Abareshi et al. [15] have evaluated the thermal conductivity of nanofluids with different volume fractions of Fe_3O_4 NPs, at different temperatures. The thermal conductivity was reported to increase up to 11.5% for a volume fraction of 3 vol% at 40°C. Xia et al. [16] have investigated the heat transfer coefficient of nanofluids using TiO_2 and Al_2O_3 NPs, with different volume fractions. For a volume fraction of 1%, the heat transfer coefficient was significantly increased for both nanofluids, compared with deionized water. Gavili et al. [17] have studied the thermal conductivity of ferrofluids with Fe_3O_4 particles of approximately 10 nm in diameter, suspended in deionized water. With the application of a magnetic field, the thermal conductivity was increased up to 200% for a 5% volume fraction. Kim et al. [18] have investigated the thermal conductivity of alumina- and distilled water-based nanofluids, with concentrations of 0.5, 1, and 2 wt%. The conductivity was found to increase with the increasing of the NPs concentration. Al-Rjoub et al. [19] have tested four cooling liquids: deionized water; distilled water; borax buffer; and Al_2O_3 NPs solution, on a microscale heat exchanger. It was found that the deionized water has presented the lowest heat removal capacity, while the Al_2O_3 solution showed the highest capacity, corresponding to about 69% increase.

The majority of microchannel heat sink devices that can be found in literature were fabricated in silicon, due to its thermal conductivity. However, the fabrication process of those devices can be laborious and needs extremely expensive facilities. Novel, fast, and low-cost fabrication techniques have been developed by means of different kinds of polymers [20–22]. PDMS is a silicone elastomer with a set of properties that make it suitable for many applications and is a popular choice for microfluidic devices fabrication [21, 22]. Besides being cheaper than the monocrystalline silicon, it presents a low elasticity change versus temperature, high thermal stability, chemical inertness, dielectric stability, shear stability, high compressibility, and hyperelasticity [23–27]. Moreover, it is non-toxic and biocompatible [25, 27–29]. PDMS devices can be manufactured by simple techniques at room temperature, such as replica molding.

The 3D printers are gaining an increased attention by both academic and industrial community to produce microdevices and models at an extremely low cost. Some successful applications can already be found in lab-on-a-chip tools [30], microfluidics [31, 32], and biomedical *in vitro* devices [27, 32–34]. There are different kinds of printing methods such as the Fused Deposition Modeling (FDM) and stereolithography [32]. Among those methods, the most popular, simple method with the lowest cost is the FDM technology [32]. For this reason, the FDM process was combined with a PDMS casting technique to produce a PDMS microfluidic device. Hence, the main objective of this work is to show the potential of a FDM 3D printer to produce microfluidic devices and their potential to be used to perform flows and heat transfer studies of nanofluids. To demonstrate the potential of this technique, fluid flow and heat transfer studies were performed by using different fluids such as distilled water-, alumina (Al_2O_3)-, and iron oxide (Fe_3O_4)-based nanofluids with concentrations of 1 and 2.5%.

2. Materials and methods

This section describes the experimental protocol to develop the nanofluids and the fabrication process to produce the heat sinks, as well as the used experimental setup.

2.1 Nanoparticles preparation

Two different types of NPs were used on the flow and heat experiments: alumina oxide (Al_2O_3) NPs acquired from Sigma Aldrich (ref. 702,129, Sigma Aldrich) and iron oxide (Fe_3O_4) NPs synthesized on our laboratory by co-precipitation. This last method allows to produce magnetic iron oxide NPs in a cost-effective way and is appropriate for mass production. The Al_2O_3 NPs had a size inferior to 50 nm, while the synthesized Fe_3O_4 NPs had an average size of 11 ± 2 nm.

The co-precipitation was initiated with the preparation of the precipitation agent by adding 0.01 g of cetrimonium bromide (CTAB), diluted in 3 mL of distilled water, to 20 mL of ammonium hydroxide (NH_4OH). Hereinafter, a ferrous solution was prepared by diluting 7.78 g of iron(III) chloride ($FeCl_3$) and 4.06 g of iron(II) chloride ($FeCl_2$) in 20 mL of distilled water, in an ultrasonic bath. The solution was subsequently mechanically stirred at, approximately, 1500 rpm. The precipitation solution was then added, dropwise, to the ferrous solution under stirring, on a laminar flow cabinet. The co-precipitation occurred according to the following equation:

$$\begin{aligned} FeCl_2.4H_2O_{(l)} + 2FeCl_3.6H_2O_{(l)} + 8NH_4(OH)_{(aq)} \\ = Fe_3O_{4(s)} + 8NH_4Cl_{(aq)} + 20H_2O_{(l)} \end{aligned} \tag{1}$$

To conclude the process, the NPs were washed several times with distilled water with the assistance of a strong magnet.

Figure 1 shows the synthesized NPs and the representative transmission electron microscopy (TEM) image. A detailed description of this process can be found at Cardoso et al. [35].

Despite the TEM images show the aggregates of Fe_3O_4 NPs, in the NF, they were stable and non-aggregated for a period of time after 10 min of ultrasonic bath.

The NPs show a normalized magnetization of ~69 emu.g 1 at ~10 kOe, which corresponds to the saturation magnetization, and also show a superparamagnetic behavior with an extremely low coercivity of 1.6 Oe [35–37].

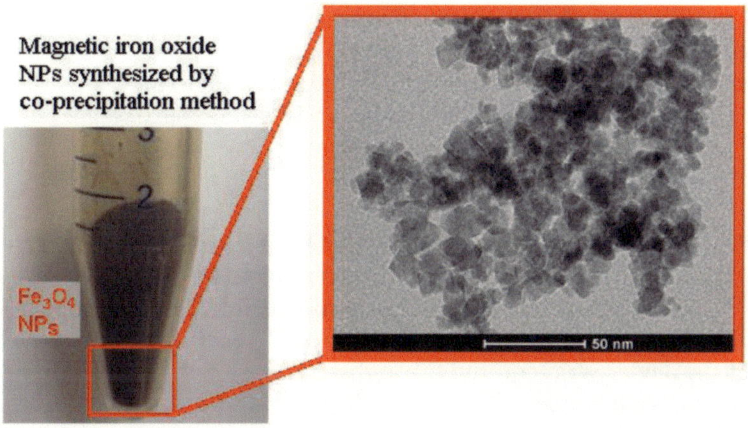

Figure 1.
Magnetic iron oxide NPs and representative TEM image, adapted from [35, 36].

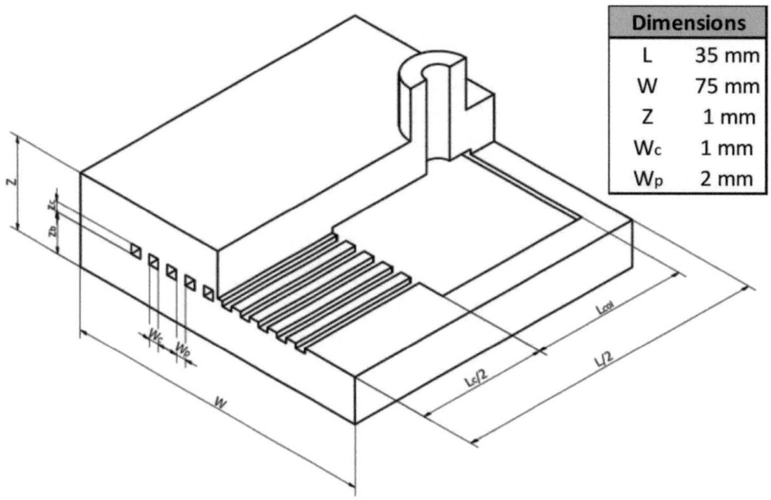

Figure 2.
Schematic representation of the main dimensions of the ABS master mold.

2.2 Fabrication of the heat sink microchannel

The heat sink microchannel device was produced based on a scaffold-removal technique [25]. First, the molds were drawn by using the Autodesk Inventor® software and then printed at the FDM 3D printer Ultimaker 2+ (Ultimaker, Netherlands). The first mold was printed with acrylonitrile butadiene styrene (ABS), whereas the second one was printed with polylactic acid (PLA). The fabrication of the molds was performed with a nozzle with a diameter of 0.4 mm, whereas the layer resolution was about 100 μm. The main dimensions of the ABS master mold can be found in **Figure 2**.

Once the 3D models were printed, PDMS was prepared by adding a PDMS curing agent into the pre-polymer with a mixing ratio of 1:10. The PDMS was poured onto the PLA mold with the ABS master mold inside it. Once the PLA mold was filled with PDMS, the elastomer was cured at room temperature for about 1 day. Finally, the PDMS was removed from the PLA mold and immersed in an acetone bath to remove the ABS for approximately 24 h. **Figure 3** shows the schematic diagram of all the main steps to produce the PDMS heat sink device. The overall cost to fabricate the PDMS heat sink device is about 3.8 €. This cost includes the printing of the ABS master mold (~1 €) and PDMS casting process (~2.8 €).

Notice that after the PDMS curing process, small holes were made below the inlet and outlet to insert the thermocouples (type K). **Figure 4** shows the PDMS heat sink microfluidic device used in the flow and heat experiments.

2.3 Experimental procedures

The PDMS heat sink was placed on top of a hot plate controlled by a 9400-temperature controller (CAL Controls). The temperature of the plate was set to 60°C, whereas the flow rate of the fluids was controlled by a syringe pump (Harvard) connected to the inlet of the heat sink. The temperature at the entrance and exit of the device was acquired through a data acquisition instrument connected to the thermocouples of the device. Wood and polystyrene blocks were used to minimize the heat losses. **Figure 5** shows a schematic diagram of the experimental setup. The flow of the Fe_3O_4 nanofluid was analyzed by optical microscopy at a flow rate of 10 μL/min. Note that the temperature measurements presented

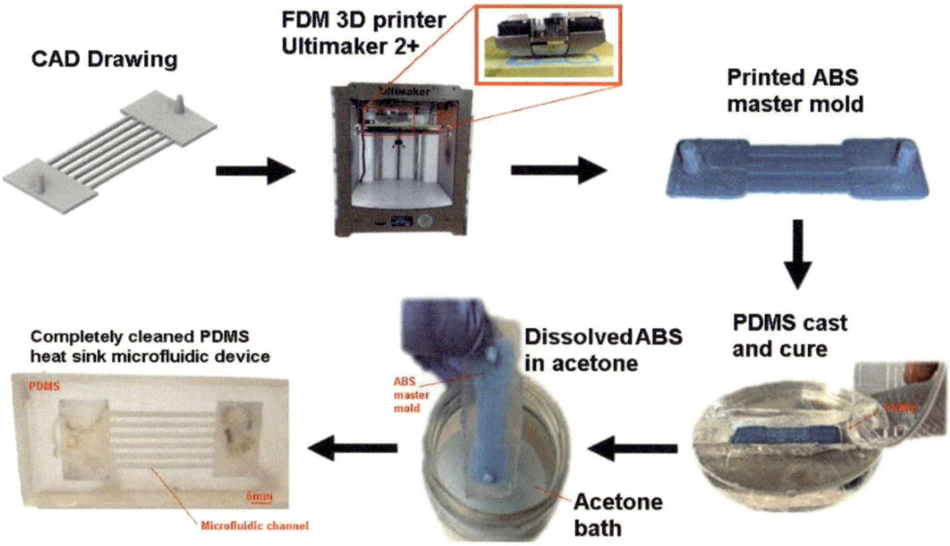

Figure 3.
Schematic representation of main steps to fabricate the PDMS heat sink.

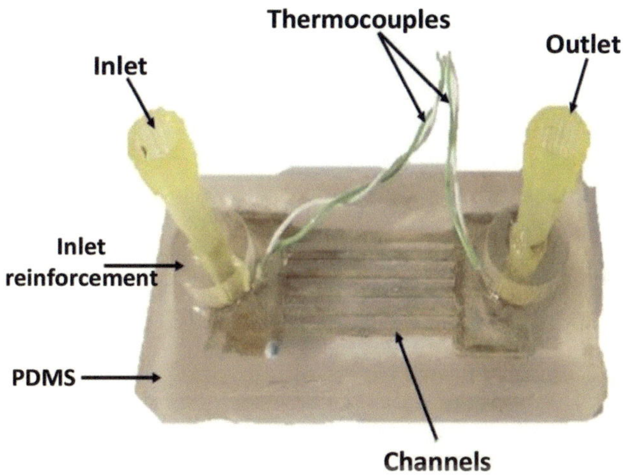

Figure 4.
PDMS heat sink microfluidic device with the inserted thermocouples.

an uncertainty of ±1°C. The thermographic studies were performed with distilled water at a flow rate of 1, 5, 7.5, 10, and 15 mL/min and a thermographic camera Onca-MWIR-InSb (Xenics Infrared Solutions). The setup and calibration procedures of the camera were performed as in Teodori et al. [38]. Images, with a resolution of 150px × 150px, were taken with a frame rate of 1000 fps.

To evaluate the influence of the nanofluids properties in the heat sink microfluid device, the tests were performed using distilled water, Fe_3O_4 at a concentration of 1 and 2.5%, and Al_2O_3 at the same concentrations. All the fluids were set to a flow rate of 1 up to 30 mL/min.

2.4 Heat transfer calculations

The properties of the nanofluids were obtained taken into account fundamental equations described on previous studies [39, 40]. The thermal conductivity of the

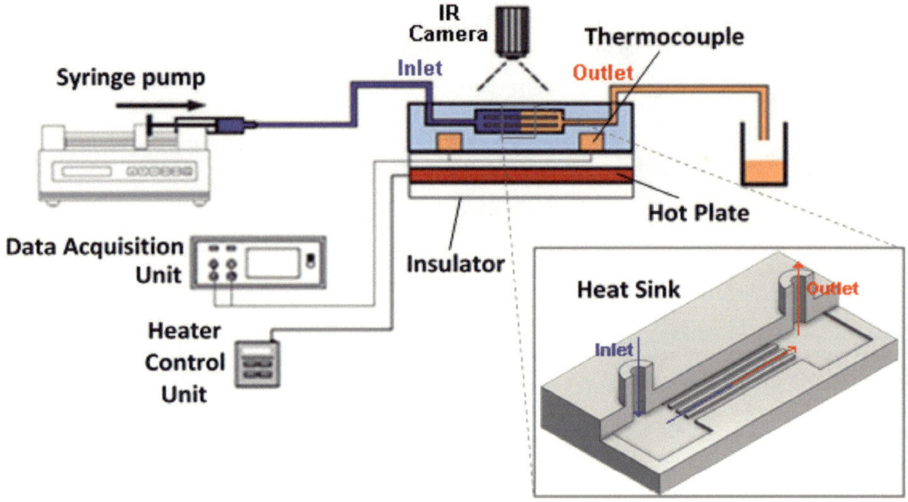

Figure 5.
Schematic representation of the experimental setup.

nanofluid was obtained according to the Maxwell model described by the following Equation [39, 40]:

$$K_{nf} = K_{bf}\left(\frac{K_p + 2K_{bf} + 2\varphi(K_p - K_{bf})}{K_p + 2K_{bf} - 2\varphi(K_p - K_{bf})}\right), \tag{2}$$

where K_{nf} is the nanofluid thermal conductivity, K_p is the NPs thermal conductivity, K_{bf} is the base fluid thermal conductivity, and φ is the NPs concentration.

The nanofluid density and heat capacity were calculated through the weighted average of the individual properties of both the NPs and base fluid. The first is expressed by Eq. (3) [14, 41] and the latter by Eq. (4) [11, 14].

$$\rho_{nf} = \rho_p \varphi + \rho_{bf}(1 - \varphi), \tag{3}$$

$$c_{p_{nf}} = \varphi \rho_p c_{p_{np}} + (1 - \varphi) \rho_{fb} c_{p_{bf}} \tag{4}$$

In the abovementioned equations, ρ_{nf} represents the nanofluid density, ρ_p the particle density, ρ_{bf} the base fluid density, and $c_{p_{np}}$ and $c_{p_{bf}}$ the specific heat capacity of the NPs and of the base fluid, respectively.

The nanofluid viscosity was determined through the equation proposed by Batchelor [42]:

$$\mu_{nf} = \left(1 + 2.5\varphi + 6.2\varphi^2\right)\mu_{fb}, \qquad 0 < \varphi < 10\% \qquad [5]$$

where μ_{nf} is the nanofluid viscosity and μ_{fb} the base fluid viscosity. This equation brings in a quadratic dependence with the volume fraction, which provides a better representation of the interaction between the particles on the fluid.

Within the microfluidic device, the heat transfer will occur by convection inside the microchannels and by conduction in the walls between them. Consequently, the mathematical approach that better allowed the evaluation of the heat transfer was described by Ma et al. [10]. The convection heat transfer coefficient was calculated by iterations using the following equations:

$$h = \frac{Q}{N(A_b + 2\eta A_l)(T_b - T_{avgf})}, \tag{5}$$

$$m_i = \sqrt{\frac{2h_{i-1}}{k_{PDMS}w_p}}, \tag{6}$$

$$\eta_i = \frac{th(m_i z_c)}{m_i z_c}, \tag{7}$$

where A_b represents the area of microchannel bottom, A_l the area of microchannel sidewall, T_{avgf} the average temperature of the fluid, η the fin efficiency, w_p the average width of fin, and z_c the height of the channel. The main dimensions of the heat sink are illustrated in **Figure 2**. The heat transfer rate, Q, represents the amount of heat energy taken by the fluid when it flows through the channels and is given by Eq. (9). This parameter was obtained for each mass flow rate, $\dot{m}$, after the temperature at the inlet, T_{in}, and at the outlet, T_{out}, were measured.

$$Q = q\,\rho_f c_{pf}(T_{out} - T_{in}), \tag{8}$$

where q, c_{p_f}, and ρ_f is the volume flow rate, specific heat capacity, and density of the working fluid, respectively.

3. Results and discussion

In this section, the obtained results are presented and discussed. The temperature measurements and the known properties of the materials were used to calculate the parameters described on the previous section. Using those parameters, the influence of the environment conditions and of the fluid properties in heat transfer was analyzed.

3.1 Influence of the nanofluid properties

As described previously, the prepared nanofluids with iron oxide (Fe_3O_4) and alumina (Al_2O_3) NPs were used to verify the thermal properties influence of the nanofluids in the heat transfer performance of the developed PDMS heat sink device.

Figure 6 shows the temperature difference between the inlet and the outlet as the tested nanofluids flow through the proposed PDMS heat sink. Overall, it is possible to conclude that both tested nanofluids present a bigger temperature difference between the inlet and the outlet in comparison with distilled water. Hence, these measurements indicate that the amount of heat energy absorbed by both nanofluids was bigger than that absorbed by the distilled water. In addition, the amount of heat absorbed by the nanofluids was found to be bigger for smaller flow rates, as shown in **Figure 6**. These results corroborate the measurements performed by Chein and Chuang [43], where they have investigated the heat performance of nanofluids with CuO NPs in a microchannel heat sink. These results also show that the flow rate affects the amount of heat absorbed by the nanofluids.

In **Figure 7**, an increase of the convective heat transfer coefficient was registered on both nanofluids, in comparison to the base fluid. The increase was more pronounced for the alumina nanofluid due to the greater stability of the nanofluid and bigger thermal conductivity of these kind of NPs [6]. By increasing the convective heat transfer coefficient, an increase of the heat transfer rate is expected. In **Figure 8**, it is possible to observe an increase in the heat transfer rate for both tested nanofluids in comparison with distilled water. From these results, it is also possible to conclude that for both nanofluids, the convective heat transfer coefficient and heat transfer rate increase with the flow rate, which agrees with the results obtained by Wen and Ding [44].

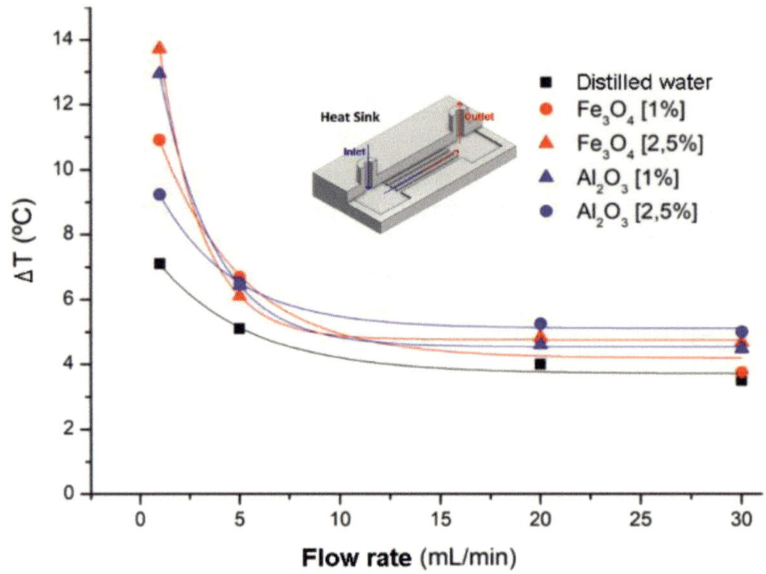

Figure 6.
Temperature difference between the inlet and the outlet for the tested nanofluids at the proposed PDMS heat sink microfluidic device.

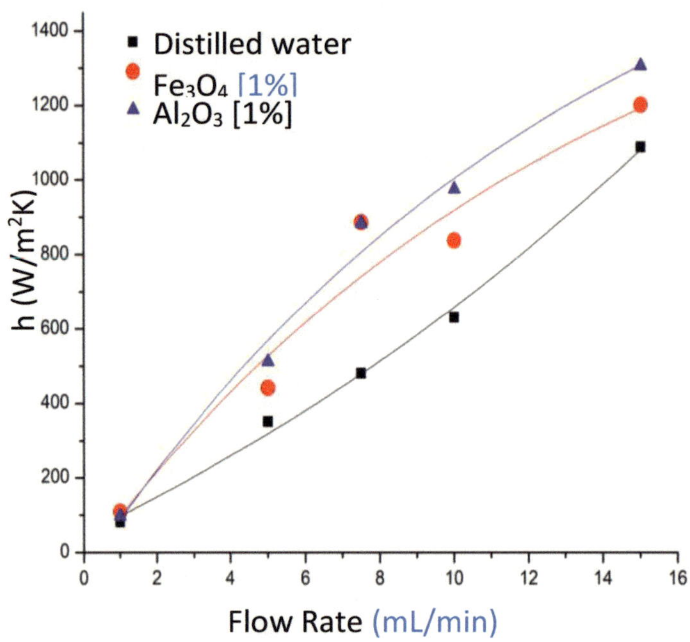

Figure 7.
Convective heat transfer coefficient of distilled water, alumina, and iron oxide nanofluids as the function of the flow rate.

Figure 9 compares the convective heat transfer coefficient for two different concentrations of NPs, i.e., 1 and 2.5% of both Fe_3O_4 and Al_2O_3. From a macroscopic view, it was noted a better dispersion for the nanofluids containing 1% of NPs. In addition, by increasing the concentration, the heat transfer was not enhanced. In fact, it was noted a decrease of the convective heat transfer coefficient when the concentration of NPs was increased to 2.5%. Although these results are somewhat

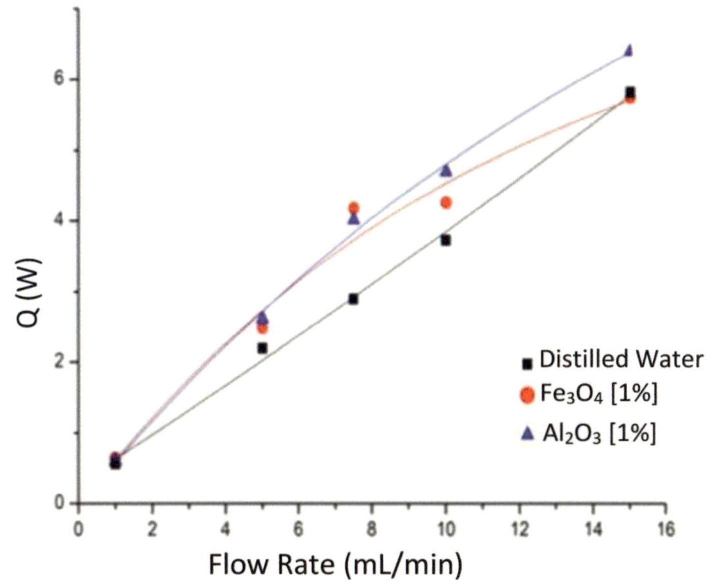

Figure 8.
Heat transfer rate of distilled water, alumina, and iron oxide nanofluids as the function of the flow rate.

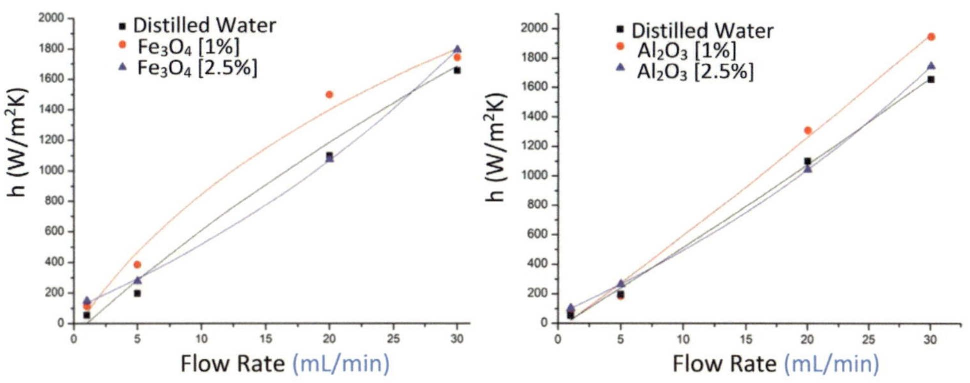

Figure 9.
Convective heat transfer coefficient of nanofluids as the function of the flow rate, for two different concentrations of alumina and iron oxide NPs.

paradoxical, other researchers, such as Wen and Ding [45] and Putra et al. [46], have reported similar results. The main possible reasons for the seen heat transfer deterioration include both the aggregation and sedimentation of the NPs. However, the possible reasons and mechanism attributed to such phenomena require further research. Currently, our group is carrying out both experimental and numerical work to identify the exact causes for such phenomena.

3.2 Optical and thermal imaging analyses

The biggest advantage of the developed PDMS heat sink device is the ability to visualize the flow phenomena happening inside the microchannels. By using a high-speed video microscopy system, it was possible to visualize several flow phenomena of the nanofluids such as the formation, growing, and breakdown of NPs clusters (see **Figure 10**). From these observations, it was concluded that one of the main causes for the formation of the clusters was the high roughness of the PDMS surface

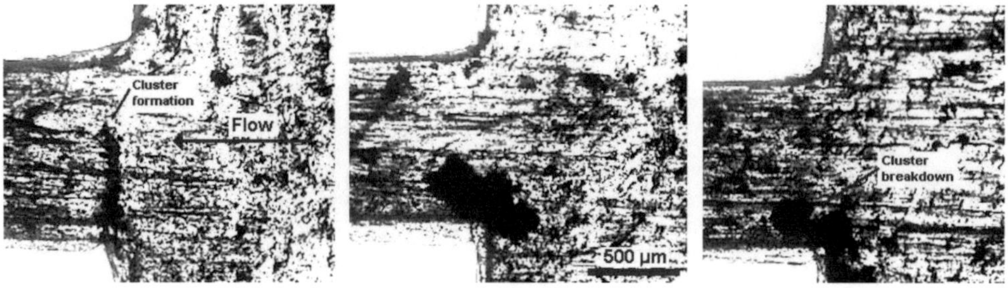

Figure 10.
Optical images of the formation, growing, and breakdown of a cluster of NPs.

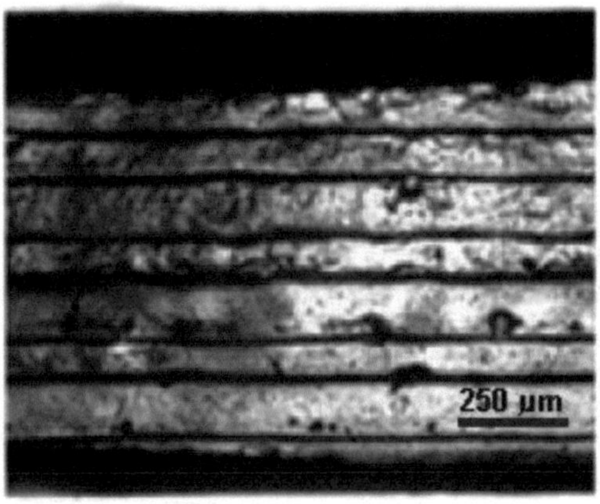

Figure 11.
Optical image of the surface roughness of the heat sink device used in this study.

channels (**Figure 11**). This verified roughness was caused by the ABS master mold fabricated by the FDM 3D printer. In order to improve the surface roughness, the ABS master molds should undergo an acetone vapor treatment before performing the PDMS casting procedure. More detailed information about this method can be found elsewhere [27].

Another interesting advantage of this PDMS microfluidic device is the ability to visualize both the flow and thermal performance of the system by using a thermographic camera, as shown in **Figure 12**. Notice that the temperatures acquired were from the surface of the heat sink device and not directly from the working fluid flowing in the microchannels. In fact, PDMS is transparent to visible radiation but partially opaque to the infrared (IR) radiation. Hence, to obtain the temperatures more closely related to the working fluids flowing through the microchannels, the thickness of the upper walls should be reduced in future experiments. A very interesting observation was the ability to detect bubbles that are likely to happen in microfluidic devices. In **Figure 13**, it is possible to visualize a bubble within the microchannel and the thermal performance of the heat sink device.

3.3 Limitations and future directions

In this study, a microchannel device was successfully manufactured and used in microfluidic essays. Nevertheless, some limitations arose throughout the work. Despite the advantages of using PDMS, some properties of the material, such as

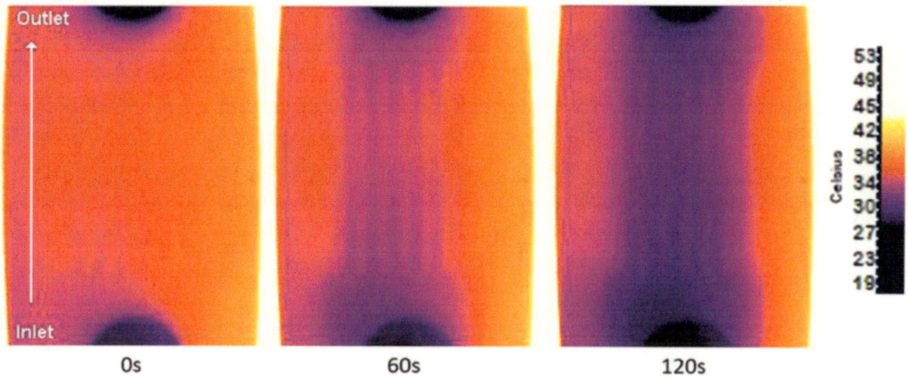

Figure 12.
Temperature gradient analyzed through the thermographic camera at three different instants: t = 0 s; t = 60 s, and t = 120 s.

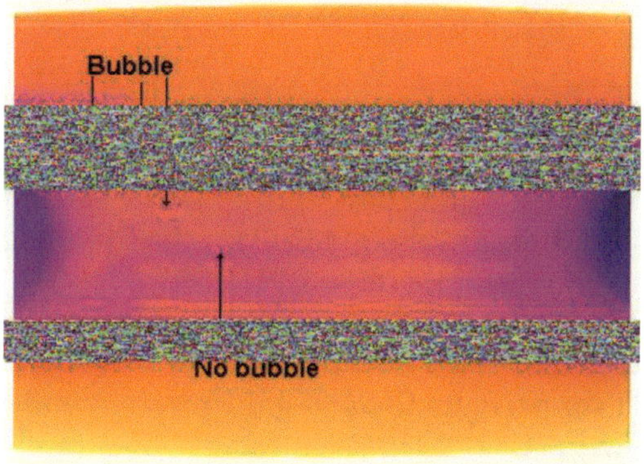

Figure 13.
Formation of a bubble at the entrance of the microchannels, which affects the thermal performance of the microfluidic device.

the low conductivity and partial opacity to the infrared (IR) radiation, were not favorable for the experiments. Also, the walls of the heat sink were rough due to the ABS master mold used in the fabrication technique, and the thickness should be reduced. The stability of the nanofluids has yet to be optimized since the deposition of NPs was detected on the heat sink walls. Future works will aim to improve of those limitations.

4. Conclusion

The main objective of this work was to show the potential of a PDMS heat sink microfluidic device to perform flows and heat transfer studies of nanofluids. The PDMS heat sink device was produced by using the FDM 3D printing process, combined with a PDMS casting technique. This fabrication process allowed to manufacture devices in an easy, low-cost, and reasonable reproductively way. To demonstrate the potential of the produced PDMS heat sink device, fluid flow and heat transfer studies were performed by using two different nanofluids, i.e., alumina (Al_2O_3)- and iron oxide (Fe_3O_4)-based nanofluids with concentrations of 1 and 2.5%. Overall, it was found that the thermal performance of the working nanofluids is in good agreement with several past studies. For instance, it was noted that the heat energy absorbed by both nanofluids was higher than that absorbed by the distilled water. In addition, it was found that the flow rate affects the amount of heat absorbed by the nanofluids. However, the most interesting and unique results were the optical and thermal imaging results. These results were only possible due to the optical transparency of the PDMS heat sink device. Hence, by using this device, it was possible to visualize several flow phenomena of the nanofluids such as the formation, growing, and breakdown of NPs clusters. From these latter observations, it was possible to conclude that one of the main causes for the formation of the clusters was the high roughness of the PDMS surface channels caused by the surface roughness of the ABS master mold fabricated by the FDM 3D printer. This drawback can be overcome by performing an acetone vapor treatment before performing the PDMS casting. Overall, the simplicity, low-cost, and unique features of the proposed PDMS heat sink microfluidic device may prove a viable alternative tool to investigate nanofluids flow and heat transfer phenomena that are not possible to be performed by the current traditional systems.

Acknowledgements

This work was supported by *Fundação para a Ciência e a Tecnologia* (FCT) under the strategic grants UID/EMS/04077/2019, UID/EEA/04436/2019, and UID/EMS/00532/2019. The authors are also grateful for the funding of FCT through the projects POCI-01-0145-FEDER-016861, POCI-01-0145-FEDER-028159, NORTE-01-0145-FEDER-029394, and NORTE-01-0145-FEDER-030171, funded by COMPETE2020, NORTE2020, PORTUGAL2020, and FEDER. The authors also acknowledge FCT for partially financing the research under the framework of the project UTAP-EXPL/CTE/0064/2017, financiado no âmbito do Projeto 5665-Parcerias Internacionais de Ciência e Tecnologia, UT Austin Programme. Ana S. Moita also acknowledges FCT for her contract int the context of the recruitment programme FCT Investigator (IF/00810/2015) and exploratory project associated with it.

Conflict of interest

The authors declare no conflict of interest.

Nomenclature

A	area
cp	heat capacity
h	convection heat transfer coefficient
K	thermal conductivity
N	number of microchannels
q	volume flow rate
Q	heat rate
R	resistance
Rt	thermic resistance
T	temperature
wp	average width of fin
zc	height of the channel

Greek symbols

η	fin efficiency
μ	viscosity
φ	nanoparticle concentration
ρ	density

Subscript

b	microchannel bottom
bf	base fluid
ext.	exterior
f	fluid
in	inlet
l	microchannel sidewall
nf	nanofluid
out	outlet
p	nanoparticles

Author details

Inês Maia[1], Cesar Rocha[2], Pedro Pontes[1], Vanessa Cardoso[3], João M. Miranda[4], Ana S. Moita[1], G. Minas[3], António L.N. Moreira[1] and Rui Lima[2,4*]

1 IN+ Center for Innovation, Technology and Policy Research, Instituto Superior Técnico, Universidade de Lisboa, Lisboa, Portugal

2 MEtRICs, Mechanical Engineering Department, University of Minho, Guimarães, Portugal

3 CMEMS-UMinho, University of Minho, Guimarães, Portugal

4 CEFT, Faculdade de Engenharia da Universidade do Porto (FEUP), Porto, Portugal

*Address all correspondence to: rl@dem.uminho.pt

References

[1] Moura M, Teodori E, Moita AS, Moreira ALN. 2 phase microprocessor cooling system with controlled pool boiling of dielectrics over micro-and-nano structured integrated heat spreaders. In: 15th IEEE Intersociety Conference on Thermal and Thermomechanical Phenomena in Electronic Systems (ITherm). 2016. pp. 378-387

[2] Abreu VTB e. Test and optimization of a two-phase thermosyphon cooling system for microprocessors under real working conditions. Instituto Superior Técnico; 2017

[3] Abreu V, Harrison M, Gess J, Moita AS. Two-phase thermosiphon cooling using integrated heat spreaders with copper microstructures. In: Proc 17th Intersoc Conf Therm Thermomechanical Phenom Electron Syst ITherm 2018. 2018. pp. 645-652

[4] Lameiras MM. Development of a Custom Made Condenser for a Two-Phase Thermosyphon CPU Cooling System. Instituto Superior Técnico; 2018

[5] Cahill DG, Braun PV, Chen G, Clarke DR, Fan S, Goodson KE, et al. Nanoscale thermal transport. II. 2003-2012. Applied Physics Reviews. 2014;**1**(1):11305-5-11305-6. Available from: http://aip.scitation.org/doi/10.1063/1.4832615

[6] Lomascolo M, Colangelo G, Milanese M, De Risi A. Review of heat transfer in nanofluids: Conductive, convective and radiative experimental results. Renewable and Sustainable Energy Reviews. 2015;**43**:1182-1198. DOI: 10.1016/j.rser.2014.11.086

[7] Godson L, Raja B, Mohan Lal D, Wongwises S. Enhancement of heat transfer using nanofluids—An overview. Renewable and Sustainable Energy Reviews. 2010;**14**(2):629-641

[8] Saidur R, Leong KY, Mohammad HA. A review on applications and challenges of nanofluids. Renewable and Sustainable Energy Reviews. 2011;**15**(3):1646-1668. DOI: 10.1016/j.rser.2010.11.035

[9] Philip J, Shima PD. Thermal properties of nanofluids. Advances in Colloid and Interface Science. 2012;**183-184**:30-45

[10] Ma DD, Xia GD, Wang J, Yang YC, Jia YT, Zong LX. An experimental study on hydrothermal performance of microchannel heat sinks with 4-ports and offset zigzag channels. Energy Conversion and Management. 2017;**152**:157-165

[11] Xuan Y, Roetzel W. Conceptions for heat transfer correlation of nanofluids. International Journal of Heat and Mass Transfer. 2000;**43**(19):3701-3707

[12] Philip J, Shima PD, Raj B. Evidence for enhanced thermal conduction through percolating structures in nanofluids. Nanotechnology [Internet]. 30 Jul 2008;**19**(30):305706 (7pp). Available from: http://stacks.iop.org/0957-4484/19/i=30/a=305706?key=crossref.3d918427472835fc73f454942400dcac

[13] Zhu H, Zhang C, Liu S, Tang Y, Yin Y. Effects of nanoparticle clustering and alignment on thermal conductivities of Fe_3O_4 aqueous nanofluids. Applied Physics Letters. 2006;**89**(2):4-7

[14] Pang C, Lee JW, Kang YT. Review on combined heat and mass transfer characteristics in nanofluids. International Journal of Thermal Sciences. 2015;**87**:49-67. DOI: 10.1016/j.ijthermalsci.2014.07.017

[15] Abareshi M, Goharshadi EK, Mojtaba Zebarjad S, Khandan Fadafan H, Youssefi A. Fabrication, characterization

and measurement of thermal conductivity of Fe_3O_4 nanofluids. Journal of Magnetism and Magnetic Materials. 2010;**322**(24):3895-3901. DOI: 10.1016/j.jmmm.2010.08.016

[16] Xia GD, Liu R, Wang J, Du M. The characteristics of convective heat transfer in microchannel heat sinks using Al_2O_3 and TiO_2 nanofluids. International Communications in Heat and Mass Transfer. 2016;**76**:256-264

[17] Gavili A, Zabihi F, Isfahani TD, Sabbaghzadeh J. The thermal conductivity of water base ferrofluids under magnetic field. Experimental Thermal and Fluid Science. 2012;**41**:94-98. DOI: 10.1016/j.expthermflusci.2012.03.016

[18] Kim S, Tserengombo B, Choi SH, Noh J, Huh S, Choi B, et al. Experimental investigation of heat transfer coefficient with Al_2O_3 nanofluid in small diameter tubes. Applied Thermal Engineering. 2019;**146**:346-355. DOI: 10.1016/j.applthermaleng.2018.10.001

[19] Al-Rjoub MF, Roy AK, Ganguli S, Banerjee RK. Enhanced heat transfer in a micro-scale heat exchanger using nano-particle laden electro-osmotic flow. International Communications in Heat and Mass Transfer. 2015;**68**:228-235. DOI: 10.1016/j.icheatmasstransfer.2015.09.009

[20] Al-Rjoub MF, Roy AK, Ganguli S, Banerjee RK. Enhanced electro-osmotic flow pump for micro-scale heat exchangers. In: ASME 2012 Third International Conference on Micro/Nanoscale Heat and Mass Transfer [Internet]. American Society of Mechanical Engineers; 2012. pp. 829-833. Available from: https://asmedigitalcollection.asme.org/MNHT/proceedings/MNHMT2012/54778/829/269200

[21] Faustino V, Catarino SO, Lima R, Minas G. Biomedical microfluidic devices by using low-cost fabrication techniques: A review. Journal of Biomechanics. 2016;**49**(11):2280-2292

[22] Catarino SO, Rodrigues RO, Pinho D, Minas G, Lima R. Blood cells separation and sorting techniques of passive microfluidic devices: From fabrication to applications. Micromachines. 2019;**10**:593. DOI: 10.3390/mi10090593

[23] Schneider F, Fellner T, Wilde J, Wallrabe U. Mechanical properties of silicones for MEMS. Journal of Micromechanics and Microengineering. 2008;**18**(6):065008 (9pp)

[24] Moreira NRP. Estudo de várias propriedades mecânicas do polidimetilsiloxano (PDMS) usado em dispositivos biomédicos. Instituto Politécnico de Bragança; 2013

[25] Saggiomo V, Velders AH. Simple 3D printed scaffold-removal method for the fabrication of intricate microfluidic devices. Advancement of Science. 2015;**2**(9):1500125

[26] Khanafer K, Duprey A, Schlicht M, Berguer R. Effects of strain rate, mixing ratio, and stress-strain definition on the mechanical behavior of the polydimethylsiloxane (PDMS) material as related to its biological applications. Biomedical Microdevices. 2009;**11**(2):503-508

[27] Pinho D, Bento D, Ribeiro J, Lima R, Vaz M. An In vitro experimental evaluation of the displacement field in an intracranial aneurysm model. In: Flores P, Viadero F, editors. New Trends in Mechanism and Machine Science. Cham: Springer International Publishing; 2015. pp. 261-268

[28] Mata A, Fleischman AJ, Roy S. Characterizationofpolydimethylsiloxane (PDMS) properties for biomedical micro/nanosystems. Biomedical Microdevices. 2005;7(4):281-293

[29] Ainla A. Valves for microfluidic devices [Internet]. Chalmers University of Technology; 2007. Available from: http://www.ims.ut.ee/~alar/microtech/Ch1_4/Ch_1_4_4.php

[30] Comina G, Suska A, Filippini D. 3D printed unibody lab-on-a-chip: Features survey and check-valves integration. Micromachines. 2015;**6**(4):437-451

[31] Gaal G, Mendes M, de Almeida TP, Piazzetta MHO, Gobbi ÂL, Riul A, et al. Simplified fabrication of integrated microfluidic devices using fused deposition modeling 3D printing. Sensors and Actuators B: Chemical. 2017;**242**: 35-40. DOI: 10.1016/j.snb.2016.10.110

[32] Faria CL, Pinho D, Santos J, Gonçalves LM, Lima R. Low cost 3D printed biomodels for biofluid mechanics applications. Journal of Mechanical Engineering and Biomechanics. 2018;**3**(1):1-7

[33] Yap YL, Tan YSE, Tan HKJ, Peh ZK, Low XY, Yeong WY, et al. 3D printed bio-models for medical applications. Rapid Prototyping Journal. 2017;**23**(2):227-235

[34] Souza A, Ribeiro J, Lima R. Manufacturing process of a brain aneurysm biomodel in PDMS using rapid prototyping. In: Lecture Notes in Computational Vision and Biomechanics. Berlin: Springer Nature; 2019

[35] Cardoso VF, Irusta S, Navascues N, Lanceros-Mendez S. Comparative study of sol-gel methods for the facile synthesis of tailored magnetic silica spheres. Materials Research Express. 2016;**3**(7):1-7. DOI: 10.1088/2053-1591/3/7/075402

[36] Lima R, Vega EJ, Cardoso VF, Minas G, Montanero JM. Magnetic PDMS microparticles for biomedical and energy applications. In: Lecture Notes in Computational Vision and Biomechanics. Berlin: Springer Nature; 2019

[37] Cardoso VF, Miranda D, Botelho G, Minas G, Lanceros-Méndez S. Highly effective clean-up of magnetic nanoparticles using microfluidic technology. Sensors and Actuators B: Chemical. 2018;**255**:2384-2391. DOI: 10.1016/j.snb.2017.08.095

[38] Teodori E, Pontes P, Moita AS, Moreira ALN. Thermographic analysis of interfacial heat transfer mechanisms on droplet/wall interactions with high temporal and spatial resolution. Experimental Thermal and Fluid Science. 2018;**96**:284-294. DOI: 10.1016/j.expthermflusci.2018.03.013

[39] Lamas B, Abreu B, Fonseca A, Martins N, Oliveira M. Critical analysis of the thermal conductivity models for CNT based nanofluids. International Journal of Thermal Sciences. 2014;**78**:65-76. DOI: 10.1016/j.ijthermalsci.2013.11.017

[40] Aybar H, Sharifpur M, Azizian MR, Mehrabi M, Meyer JP. A review of thermal conductivity models for nanofluids. Heat Transfer Engineering. 2015;**36**(13):1085-1110

[41] Pak BC, Cho YI. Hydrodynamic and heat transfer study of dispersed fluids with submicron metallic oxide particles. Experimental Heat Transfer. 1998;**11**(2):151-170

[42] Pavlik M. The dependence of suspension viscosity on particle size, shear rate, and solvent viscosity. 2009. Available from: http://via.library.depaul.edu/etd/71

[43] Chein R, Chuang J. Experimental microchannel heat sink performance studies using nanofluids. International Journal of Thermal Sciences. 2007;**46**(1):57-66

[44] Wen D, Ding Y. Experimental investigation into convective heat transfer of nanofluids at the entrance region under laminar flow conditions.

International Journal of Heat and Mass Transfer. 2004;**47**(24):5181-5188

[45] Wen D, Ding Y. Formulation of nanofluids for natural convective heat transfer applications. International Journal of Heat and Fluid Flow. 2005;**26**(6):855-864

[46] Putra N, Roetzel W, Das SK. Natural convection of nano-fluids. Heat and Mass Transfer. 2003;**39**(8-9):775-784

Chapter 9

The Characteristics and Application of Nanofluids in MQL and MQCL for Sustainable Cutting Processes

Tran The Long and Tran Minh Duc

Abstract

Recently, there has been growing attention to nanofluids, especially in industry. More and more people nowadays see nanoparticle applications in various fields such as automotive, agriculture, medicine, machining, and so on. The addition of different nanoparticles to fluids has shown enormous advantages, particularly for improving the efficiency and therefore lowering the energy consumption of processes for addressing a wide range of global challenges related with energy and environmental problems. Nanoparticles are of great scientific interest as they are in that nanofluid with unusual effects, and ultra-small sizes will be a new area for researchers and definitely offer novel mechanisms and technologies in the future. In this chapter, the authors will mainly present the characteristics as well as latest advances in applications of nanofluids in machining practices. Nanoparticle additives contribute to reduce friction coefficient, lower the energy consumption, and significantly extend tool life by lowering thermal stress, from which the surface quality of manufactured parts improves. Moreover, the nanoparticle application in some of the green technologies as MQL and MQCL using vegetable oils not only brings out superior cooling and lubricating properties and minimizes the use of cutting fluids, but also creates new solutions for machining, especially for difficult-to-cut materials.

Keywords: nanoparticles, nanofluid, sustainable cutting, MQL, MQCL, hard machining, vegetable oil, metal cutting

1. Introduction

Climate change has become the most growing concern of people around the world. The rapid increase of population consequently leads to the use of more natural resources and giving out of more waste. The pollution in air, water, and food causes many serious human diseases. Accordingly, environmental laws are continuously tightening up to protect our Earth. Being a part of the production chain, manufacturing engineers are demanded not only to produce the products to meet the growing demand for higher quality and productivity but also to be responsible for achieving the sustainability in manufacturing. In metal cutting industries, the used cutting fluids after using for cooling and lubricating the contact zone contribute the largest amount of disposal (around 30%), which finally ends up as the contamination in the rivers leading to the water pollution [1]. Therefore, it is necessary to find the solutions to reduce or eliminate the usage of coolants. Over some last decades, there were numerous studies concerning the reduction of coolant usage in machining, and dry cutting processes, the truly environmental-friendly method, had drawn most attention and brought out the obvious cost benefits derived from the elimination and treatment of cutting fluids. However, the selection of the proper cutting tools or inserts plays a very important role to ensure the proper tool life and high precision and accuracy of machined parts [2], and it also causes a strong influence on technological and economic characteristics. Recently, to meet the continuously increasing demand for cutting difficult-to-cut materials having high-graded mechanical properties and high hardness, the tools with geometrically defined cutting edges are directly used for machining the heat-treated materials, with the typical hardness of 45–70 HRC [3]. These processes are called hard machining, which has become the research trend in mechanical applications due to high productivity and accuracy. Up to now, people have seen hard machining more in metal cutting field, and therefore many of traditional grinding processes have been replaced. The new approach not only provides the alternative solution for cutting hard materials but also improves the cutting performance, significantly reduces coolant usage, and has low machine tool investment. On the other hand, the thermal shock caused by the use of cutting fluids must be seriously considered to avoid the insert breakage, so the flood cooling is not usually used for hard machining processes, especially for interrupted cutting. Furthermore, the enormous heat and high forces arising from cutting zone are the most challenging problems of hard cutting processes, which always demand the appropriate uses of high-graded cutting tools like coated carbide, ceramic, polycrystalline cubic boron nitride (PCBN), and diamond inserts [2, 4, 5]. Accordingly, minimum quantity lubrication (MQL) technique was proposed and proven to use and exhibited the promising results in some last decades [1, 6, 7]. The cutting fluids in forms of oil mist are directly sprayed to cutting zone, so the lubricating effect is very high to decrease the friction coefficient, from which cutting forces, cutting temperature, and tool wear reduce significantly, and tool life is extended. Interestingly, the minimal use of cutting fluid makes MQL an environmental friendly technique, and the vegetable oils can be used for hard cutting, which contributes to protect environment [8]. The main drawback of MQL method is the low cooling effect, which limits the applicability and cutting performance of hard machining [9, 10]. In order to develop MQL technique, there have been many studies proposing the very promising solutions to enhance the cooling performance, which includes MQL using nanofluids, minimum quantity cooling lubrication (MQCL), and MQCL using nano additives. In this chapter, the authors mainly discuss the latest studies on those up-to-date techniques used in hard machining processes.

This chapter is divided into five sections. Section 1 of the chapter provides the literature review of new development of MQL and MQCL technology using nanofluids for sustainable cutting processes. Section 2 is dedicated to hard machining under MQL condition using nanofluid. Section 3 describes the application of MQCL condition based on the new approach for hard cutting processes. Section 4 contains the latest advances on the utilization of nano additives for improving MQCL hard machining performance. Finally, Section 5 draws out the conclusions and some suggestion for future work.

2. Hard machining under MQL condition using nanofluid

Using nano additives suspended in MQL based fluids has opened a new approach for machining difficult-to-cut materials and is also an up-to-date research topic gaining the growing concerns, especially for encountering climate change. There are many types of nanoparticles, such as Al_2O_3, MoS_2, SiO_2, ZrO_2, CuO, TiO_2, CNT, ND, and so on, proven to use for improving the tribological property, thermal conductivity, and viscosity [11].

2.1 The improvement of cutting performance

In order to apply this technique in machining practice, the parameters of MQL (the based fluid, air pressure, flow rate) and nanofluid (the type, size, and concentration of nanoparticles) are needed to study and optimize, because they have strong effects on the cutting process. If the inappropriate values of each parameter are chosen, the little effectiveness and even the negative influence may occur in machining responses.

Li et al. [12] investigated MQL grinding process for Ni-based alloy using six different types of nanofluids. The results indicated that the viscosity and thermal conductivity of nanofluids significantly improve when compared to the base fluids. The authors also pointed out that CNT nano additives exhibit the highest heat transfer coefficient. Hence, the cutting temperature and forces decrease. Another observation done by Ali et al. [13] indicated that the viscosity of Al_2O_3 and TiO_2 nano-lubricants increases while their kinematic viscosity decreases slightly. Through the experiments, the coefficient of friction, power consumption, and wear rate much reduced due to the rolling performance together with the formation of tribo-films created by Al_2O_3 nanoparticles. Moreover, the technical specification and concentration of nanoparticles play a very important role and strongly influence on the machining responses. For finish cutting, the nanoparticles with smaller grain size and higher concentration should be used to improve the surface quality and reduce the cutting forces [11]. Pashmforoush and his co-authors [14] reported that the big improvement in surface roughness in grinding process of Inconel 738 super alloy is about 62.16 and 36.36% compared to those of dry and flood conditions, respectively. The enhancement of lubricating effect was also reported in milling under MQL using MoS_2 nanofluid, from which the friction coefficient reduced to extend the tool life and improve the surface quality [15]. The presence of nanoparticles in MQL based fluid not only improves the cooling and lubricating effects but also brings out the better cutting performance for machining difficult-to-cut materials. Moreover, this approach will successfully replace the dry and flood conditions, which fulfill the technological, economic, and environmental requirements, suitable for modern manufacturing. The performance investigation of end milling of SKD 11 steel using HSS tools under nanofluid MQL was done in [16]. The experiments are set up and shown in **Figure 1**. The cutting condition includes the three values of cutting speeds of 18, 24, and 30 m/min, feed rate of 0.01 mm/tooth, and axial depth of cut of 3 mm. The diameter of end mill is 10 mm. Al_2O_3 nano additives (0.5 wt%.) are enriched in emulsion and soybean-based oil.

Through experimental results, the cutting force components F_x, F_y, and F_z when changing the cutting speeds and based fluid are shown in **Figures 2–4**. Under NFMQL with emulsion-based oil, the cutting forces F_x, F_y, and F_z reduce with the increase of cutting speed from 18 to 30 m/min. At a cutting speed of 30 m/min, the comparison between emulsion and soybean oil is made to find out the effect of the based fluid on machining responses. It can be clearly seen that the cutting forces reduce due to the better lubricating performance of soybean oil compared to emulsion fluid.

In addition, the tool life under NFMQL using soybean oil much improves and is over two times longer than that under the case using emulsion fluid (**Figure 5**). The investigation of tool wear is shown in **Figures 6–8**. They clearly reveal that notch wear and flank wear on HSS end mills increase with the rise of cutting

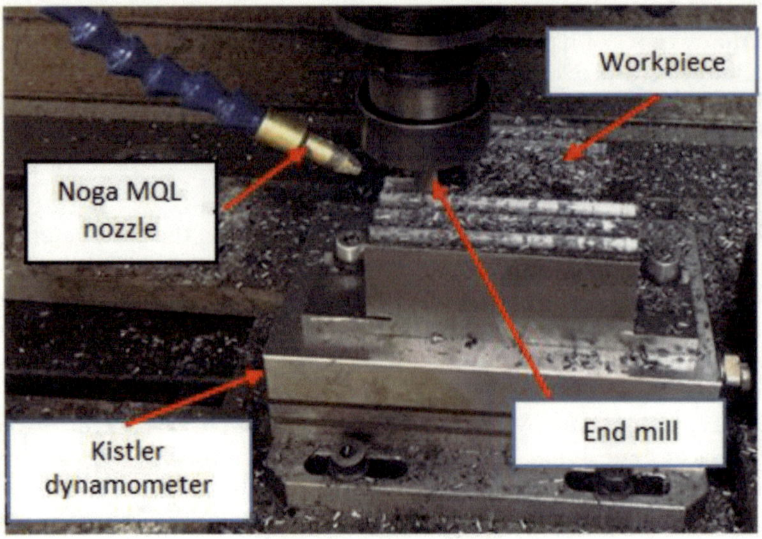

Figure 1.
Experimental setup [16].

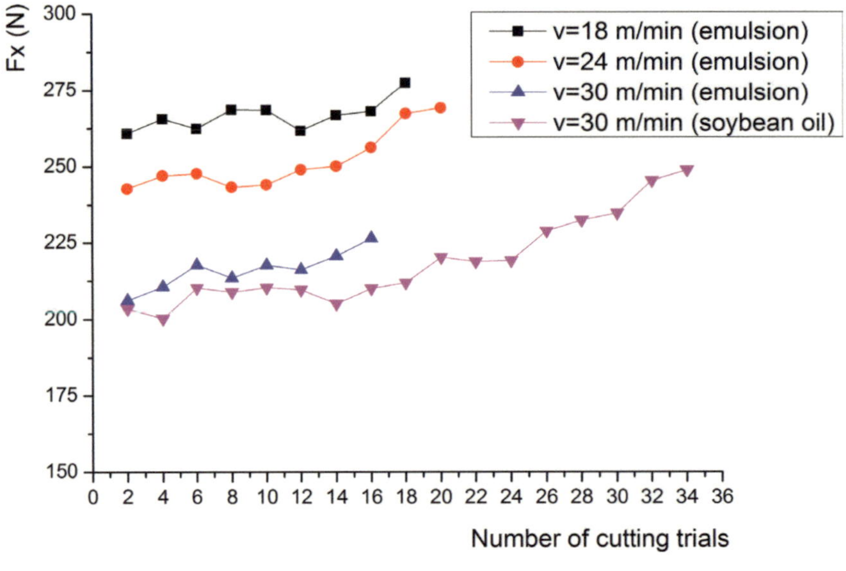

Figure 2.
The relation of cutting speeds and nanofluids to the cutting force F_x [16].

speeds from 18 to 30 m/min under MQL using emulsion. In addition to that, the burn marks caused by heat deterioration develop. It can be explained that SKD 11 tool steel has extremely high wear-resistant properties due to a high carbon and chromium (12% chrome) in chemical composition, from which it is grouped in difficult-to-cut material. In increasing the cutting speed from 18 to 30 m/min, MQL with emulsion-based fluid did not provide sufficient lubricating effects, so cutting temperature increased rapidly to damage end mills. In contrast, soybean oil has higher viscosity than that of emulsion, and the presence of Al_2O_3 nanoparticle additives contributes to improve the cooling and lubricating performance. Therefore, soybean oil-based nanofluid exhibits the superior lubricating effects to reduce friction coefficient in the cutting zone due to the easier formation of oil

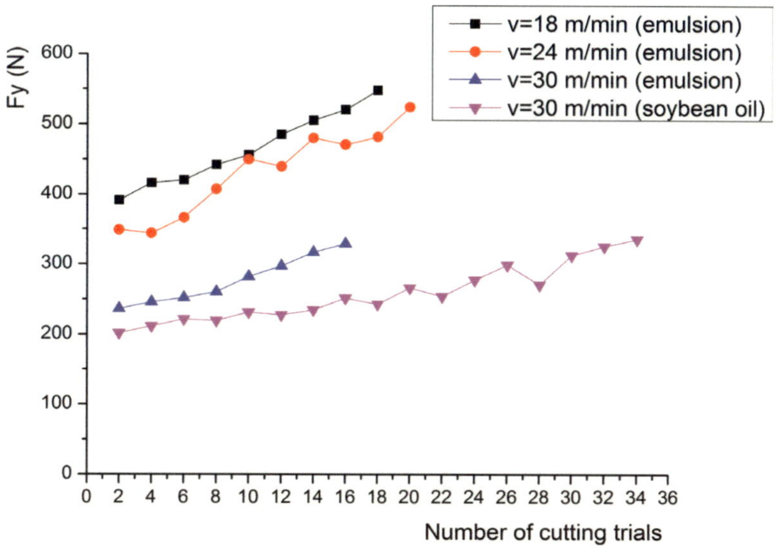

Figure 3.
The relation of cutting speeds and nanofluids to the cutting force F_y [16].

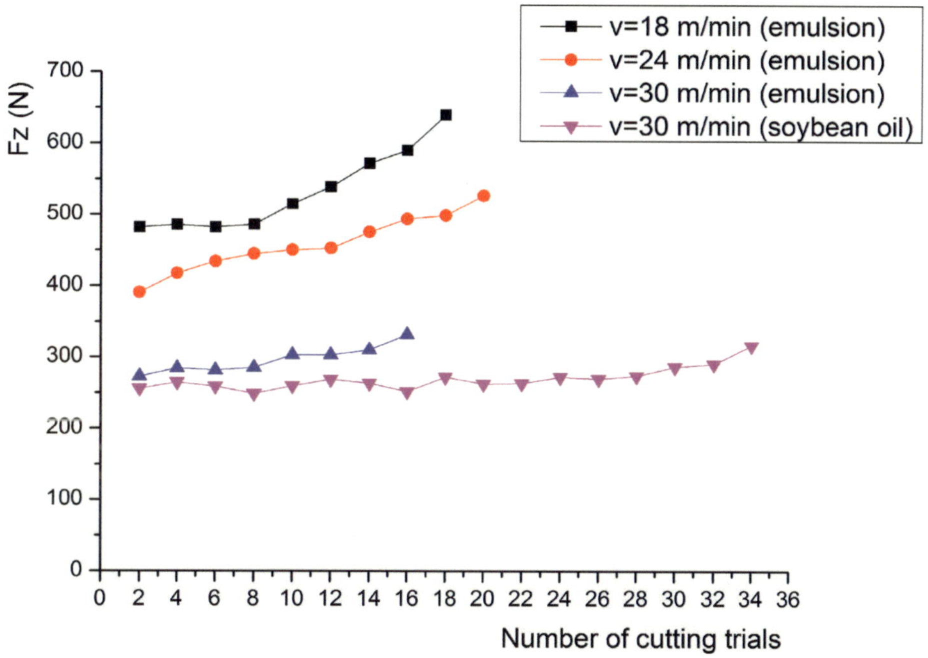

Figure 4.
The relation of cutting speeds and nanofluids to the cutting force F_z [16].

mist. In this study of end milling process of SKD 11 steel before heat treatment, cutting heat did not exceed the ignition temperature of soybean oil. That is the main reason why MQL using soybean-based nanofluid is better in this situation. Furthermore, Al_2O_3 nanoparticles with nearly sphere morphology suspended in oil mist as "the rollers" play an important role in improving cooling and lubricating effects. From those reasons, notch wear and flank wear on HSS end mills at cutting speed V_c = 30 m/min significantly reduce. Notch and flank wear lands after 85 min

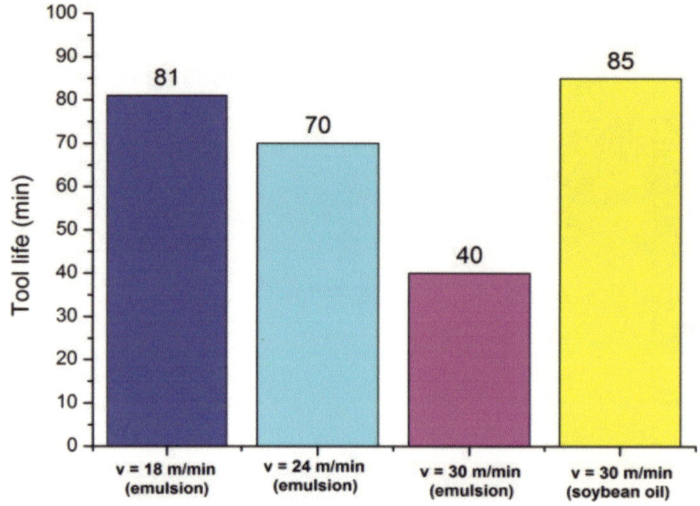

Figure 5.
The relation of cutting speeds and nano-cutting fluids to the tool life [16].

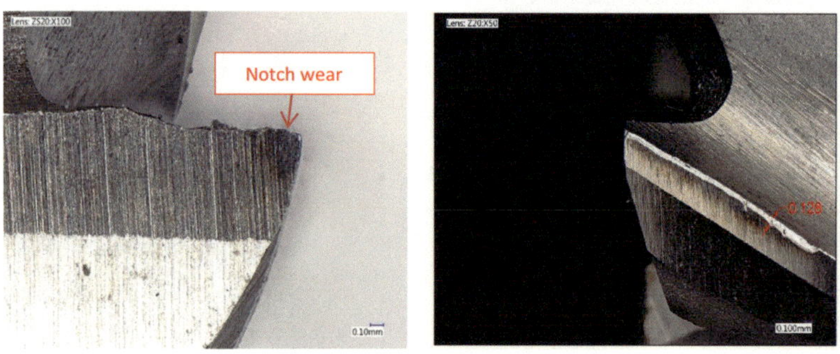

Figure 6.
Notch wear and flank wear (Vc = 18 m/min, emulsion-based nanofluid) [16].

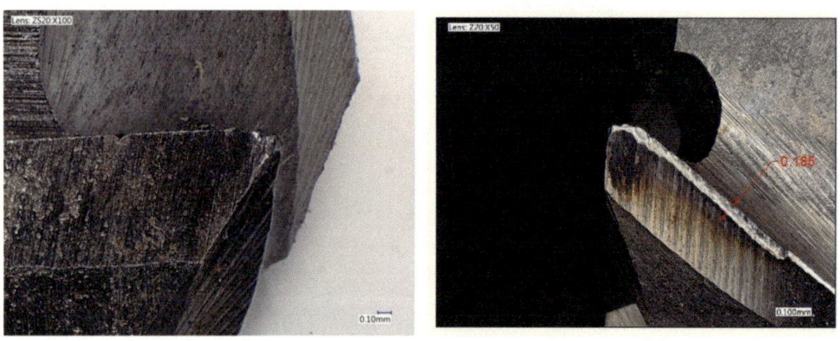

Figure 7.
Notch wear and flank wear (Vc = 24 m/min, emulsion-based nanofluid) [16].

of cutting are even lower than those of the case with MQL using emulsion-based nanofluid after 40 min of cutting (**Figures 8–9**). The significant reduction of burn marks indicates that the cooling and lubricating performance of soybean-based nanofluid is better and also suits for sustainable production due to the use of vegetable oil. Hence, tool life of end mill increases to 85 min even at cutting speed of 30 m/min, which is also higher than the manufacturers' recommendations [16]. According to ISO 8688-2:1989 (en) [17], the cutting speed for soft steels using

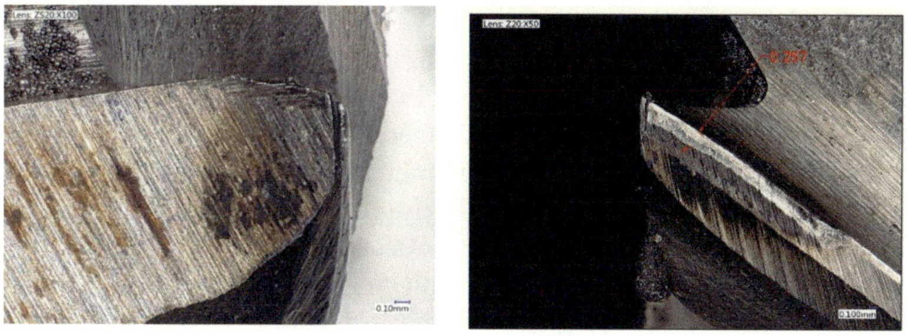

Figure 8.
Notch wear and flank wear (Vc = 30 m/min, emulsion-based nanofluid) at 40 min [16].

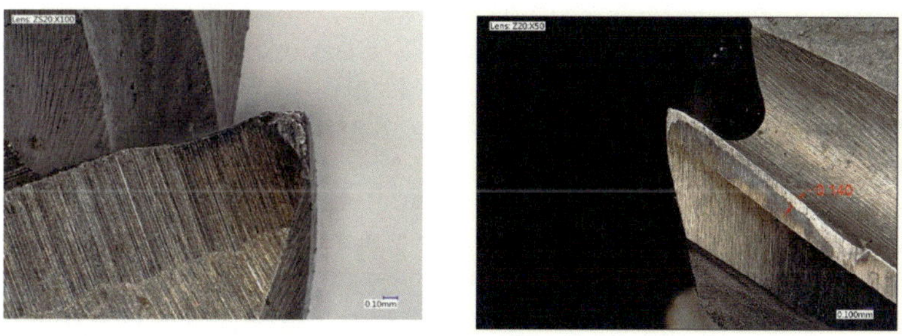

Figure 9.
Notch wear and flank wear (Vc = 30 m/min, soybean-based nanofluid) at 85 min [16].

normal HSS end mills is recommended about 30–35 m/min, but for difficult-to-cut steels like SKD 11, with hardness 200–250 HB, the cutting speed must be reduced to 14–18 m/min to ensure the proper cutting performance and tool life [18]. Moreover, the cutting speed also increases from 18 to 30 m/min by using MQL emulsion-based nanofluid, which reveals the better cooling and lubricating effects compared to the pure fluids.

2.2 The important parameter of MQL nanofluid

The concentration parameter of nanoparticles enriched in based fluids is among the most influential on machining outputs and costs, so it had been much studied to find out the appropriate and optimized values. Garg et al. [19] investigated the concentration effect of nanoparticles on micro-drilling process under MQL condition. The experimental results indicated that this parameter caused the significant reduction of drilling torque and power consumption. In the study of Lee et al. [20], the proper concentration of diamond nanoparticles was found with 0.05 wt%, from which the reduction of friction coefficient was observed by 23%. The authors concluded that diamond nanofluid provided the excellent anti-wear and lubricating effects. Zhang et al. [21] studied the concentration parameter of MoS_2 and CNT nano additives in MQL grinding. The improvement in lubricating performance contributes to increase surface quality. Furthermore, hybrid MoS_2-CNT nanofluids provided the superior cooling lubrication compared to that of the fluid with a single type of nanoparticles. Luo et al. [22] studied Al_2O_3 nanoparticles enriched in MQL based fluid and concluded that Al_2O_3 nanofluid exhibited good resistant ability for high temperature. Then, the cutting temperature is not high to cause the reduction of wear rate, which is much smaller than that of dry condition. Yıldırım et al. [23] had done the study of MQL turning process of the difficult-to-cut steel Inconel 625 using hBN nano additives. The better lubricating performance and surface roughness are reported from the obtained results, which led to reduce friction coefficient and wear rate. The authors also concluded that the optimal hBN nanoparticle concentration was 0.5 wt%. The experimental study on Al_2O_3 nanoparticle concentration used as MQL based fluid in hard milling had been done by using ANOVA analysis and response surface methodology (RSM), from which the research direction was made for optimizing the concentration variable [24]. **Figures 10–11** show the response surface plots of the relation of surface roughness and cutting force versus nano concentration (*np*), cutting speed V_c, and feed rate F. It can be clearly observed from **Figure 10** that, for better surface roughness, the low value of nanoparticle concentration about 0.5 wt% is more preferable than the larger ones (1.0 and 1.5 wt%). In contrast, the larger concentration (about 1.0 and 1.5 wt%) contributes to reduce the cutting forces and cutting temperature significantly when compared to the lower one (0.5 wt%.). From those, the wear rate much reduces by increasing the concentration of Al_2O_3 nanoparticles to 1.0–1.5 wt%, so the tool life prolongs (**Figures 12–15**). Accordingly, the nanoparticle concentration must be chosen not only to ensure the good tool life but also to maintain the high surface quality.

From **Figure 13**, the chip colors are well matched with the reduction of cutting temperature. The dark purple and blue colors in the case of using np = 0.5 wt% change to brown and dark straw one in the case of using np = 1.0–1.5 wt%, which indicates that cutting temperature decreases [25]. It can be concluded that the decrease of coefficient of friction and generated heat is reported by increasing

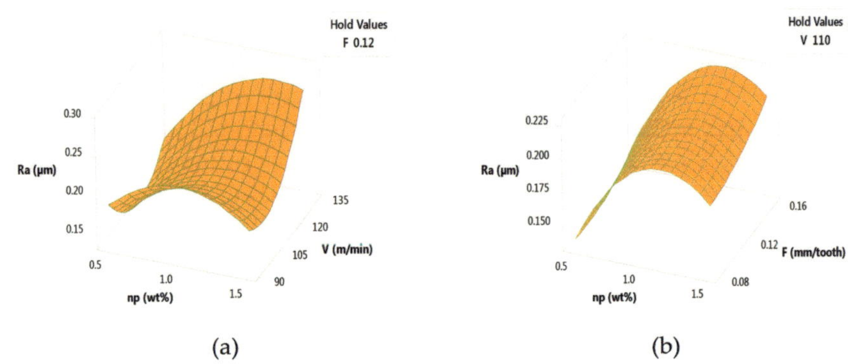

Figure 10.
Response surface plots of surface roughness versus nano concentration and cutting speed (a), and nano concentration and feed rate (b) [24].

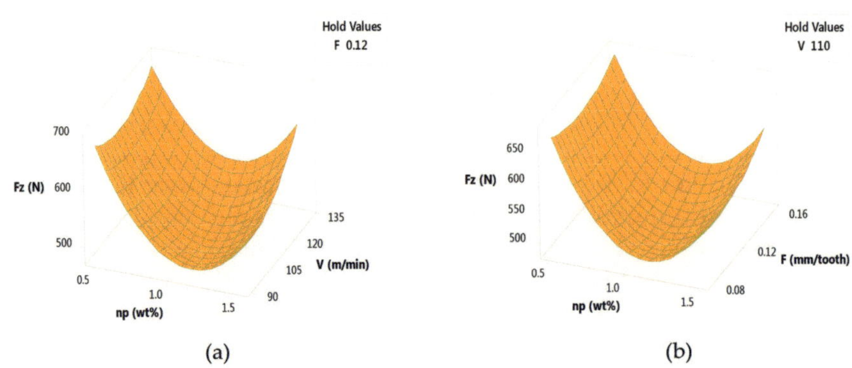

Figure 11.
Response surface plots of cutting force F_z versus nano concentration and cutting speed (a), and nano concentration and feed rate (b) [24].

the concentration parameter of Al_2O_3 nanofluid, leading to reduce cutting forces and wear rate and prolong the tool life (**Figure 14**). From **Figure 15**, it clearly reveals that during the first 40 min, the values of surface roughness are higher when utilizing the high nanoparticle concentration (1.0–1.5 wt%). After that, the

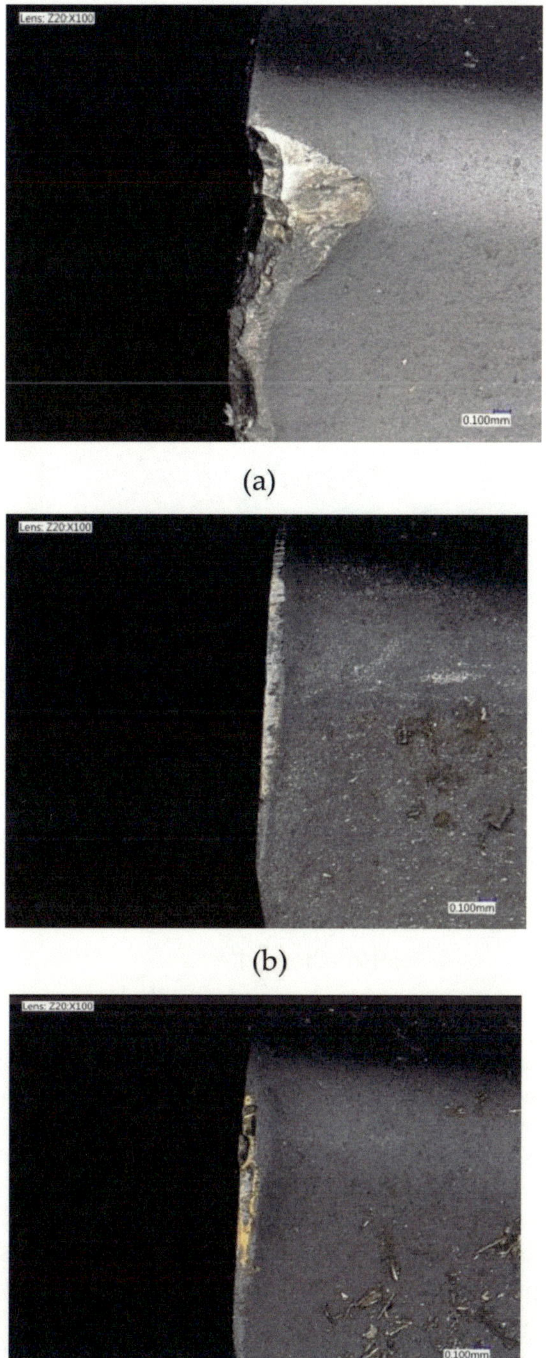

(a)

(b)

(c)

Figure 12.
Wear on flank face under MQL using soybean-based nanofluid at 80 min using different nano concentrations: (a) 0.5 wt%, (b) 1.0 wt%, and (c) 1.5 wt% [24].

Figure 13.
Chip colors and micrographs with different Al_2O_3 nanoparticle concentrations (at 80 min): (a) 0.5 wt%, (b) 1.0 wt%, and (c) 1.5 wt% [24].

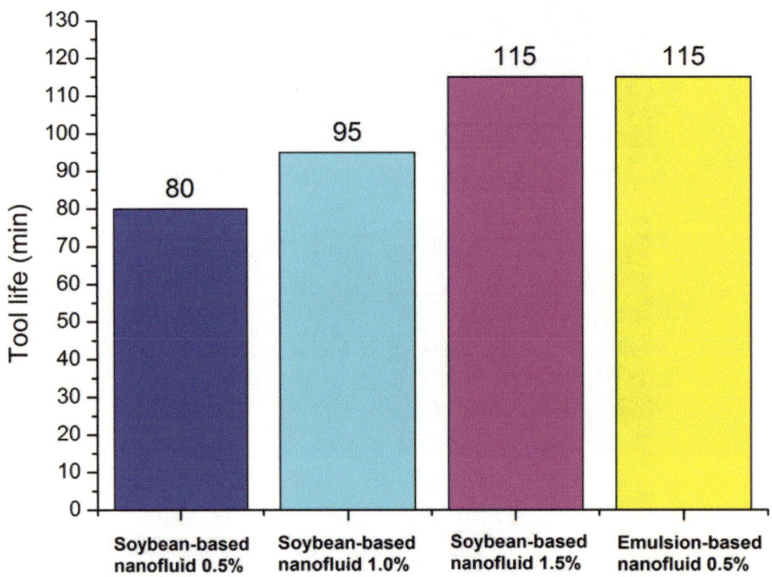

Figure 14.
Tool life with different Al_2O_3 nanoparticle concentrations [24].

rate of reduction of surface roughness values rapidly increases, which is contrary to the case of using 0.5 wt%. It is the novel observation obtained from the validation experiments, which is conducted until the tool life ends to see the actual phenomena after receiving the ANOVA and RSM results. Interestingly, the tool life in the case of soybean-based nanofluid 1.5 wt% is equal to that of emulsion-based nanofluid 0.5 wt%. It provides an important technical guide to enlarge the applicability of vegetable oil-based nanofluid in hard cutting processes as well as maintains its environmental-friendly characteristics.

However, each type of nanoparticles has its own specific morphology and property, so the appropriate or optimal concentration parameters are different [26]. This is the up-to-date research topic. Accordingly, more investigations are needed to make and build up the technical guides for manufacturers in machining practice, even though many studies have been done to optimize these variables. In order to develop MQL method, minimum quantity cooling lubrication (MQCL) has been considered as another promising approach to solve the low cooling performance, the main MQL drawback. It is also the newest research topic, which is discussed in Sections 3 and 4.

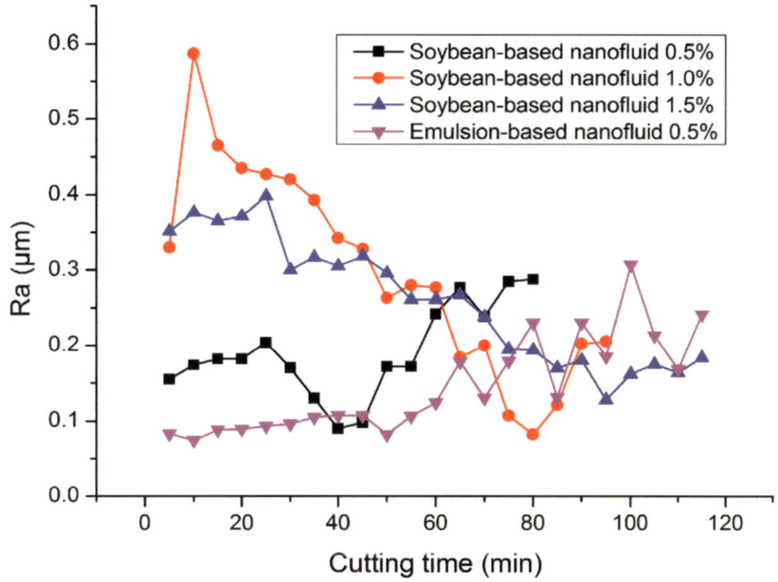

Figure 15.
Surface roughness Ra *with different Al_2O_3 nanoparticle concentrations [24].*

3. Hard machining under MQCL condition

The large amount of heat generated from hard cutting is always the big challenge for selecting cutting tools and cutting condition while ensuring the technical requirements, productivity, and proper manufacturing cost. Hence, providing appropriate cooling and lubricating effects to cutting zone plays a vital role in the development of hard machining processes. The applicability of those can be enlarged to some or all of the traditional grinding processes. MQCL technique has been proposed and developed to fulfill the cooling and lubricating requirements and is also a solution for improving MQL method. Up to now, MQCL has drawn much attention and has been studied in recent years. Maruda et al. [27, 28] made the study on MQCL parameters using emulsion-based fluid in hard turning process. The obtained results indicated that emulsion oil mist formed under MQCL condition plays an important role for improving the cooling lubricating performance in the cutting zone and increasing the cutting condition. The formation of tribo-films tends to occur easily with the droplets with smaller size, which help to decrease the coefficient of friction, cutting forces, and wear rate [29, 30]. The better cooling and lubricating effects of MQCL technique also reflect through the chip shape and the reduction of chip thickening coefficient [31]. Pervaiz and his co-authors [32] studied MQCL performance in the turning process of difficult-to-cut material Ti6Al4V. The author concluded that cutting forces and tool wear reduced and surface quality improved when compared to dry and flood conditions. It reveals the better cooling and lubricating effects of MQCL technique. Krolczyk together with his co-authors [33] investigated the parametric and nonparametric description of the surface topography under dry and MQCL conditions using emulsion-base fluid. The study results showed that the nozzle distance causes the strongest influence on droplet diameter. The most outstanding finding of this research is that parameters can be chosen for oil mist formation in a certain time, which is enough for creating cooling and lubricating effects and then evaporating due to generated heat from the cutting zone. Based on a brief review, it can be clearly seen that there is little information of MQCL technique and most of the studies relied on the based fluid having cooling effect like emulsion oil to form MQCL method. The use of a real cooling method assisted to MQL technique to form MQCL condition is a novel approach. In this section, the author presents the newest advances in using the principle of Ranque-Hilsch vortex tube for separating a compressed gas into hot and cold streams from ordinary air [34], in that the cold stream is used to create cooling effects combined with MQL method to form MQCL [35]. The deep study on hard milling of SKD 11 steel (52–60 HRC) in terms of surface quality under MQCL condition was done, and the results were compared to dry and MQL conditions. From **Figure 16**, hard milling under MQCL method brought out better surface roughness than those under dry and MQL conditions. The main reason is that MQCL technique provides sufficient cooling and lubricating effects, especially cooling effect, which helps to reduce the cutting temperature and tool wear.

KEYENCE VHX-6000 Digital Microscope (Keyence Corporation, Osaka, Japan) was utilized for studying surface microstructure and surface profile (**Figure 17**). The

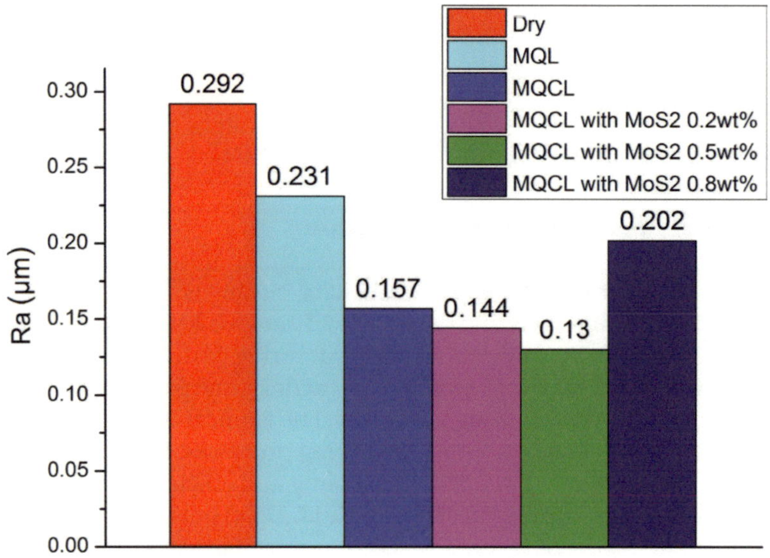

Figure 16.
The average values of surface roughness Ra under different cooling and lubricating conditions (cutting speed Vc = 110 m/min, feed rate F = 0.012 mm/tooth, depth of cut d = 0.12 mm, hardness of 56 HRC) [35].

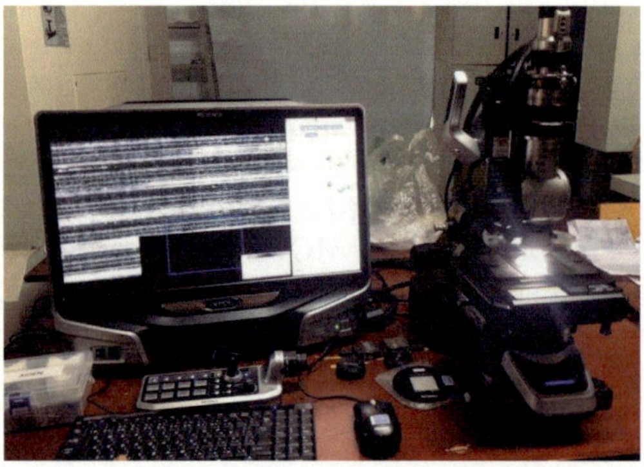

Figure 17.
KEYENCE VHX-6000 digital microscope for studying machined surface topography.

machined surfaces under different cooling and lubricating conditions are investigated (Figures **18–22**). The white layer and burn marks significantly reduced under MQL and MQCL conditions compared to dry cutting because of cooling and lubricating enhancement. The burn marks under MQCL condition are less than those under MQL method due to better cooling performance (**Figures 19(a), 20(a)**). In addition, compared to dry and MQL conditions, the compression of machined surface observed from the surface profile much reduces (**Figures 18(b), 19(b), 20(b)**).

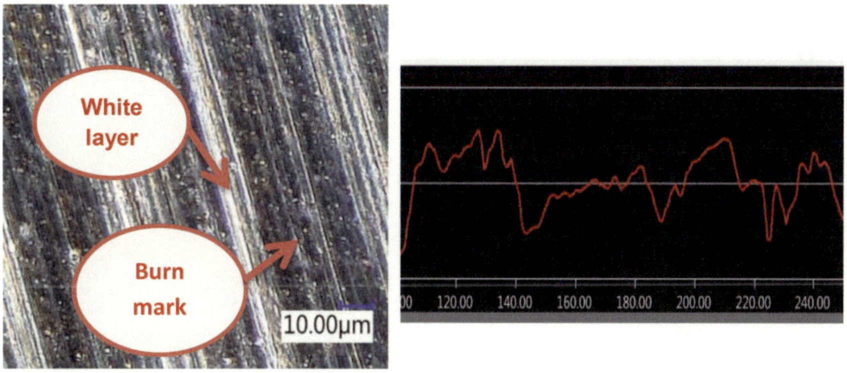

(a) (b)

Figure 18.
Surface microstructure (a) and profile (b) under dry condition [35].

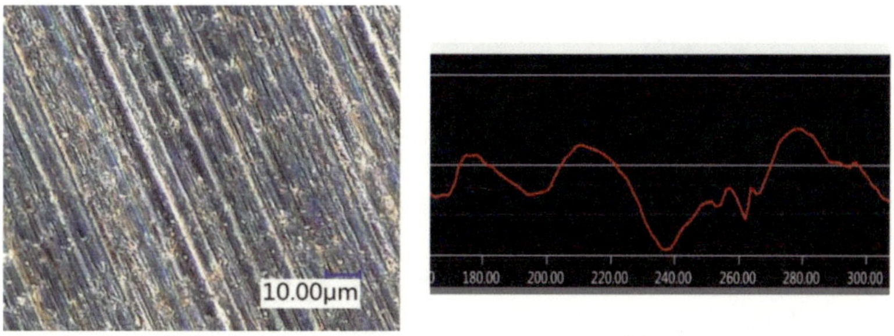

(a) (b)

Figure 19.
Surface microstructure (a) and profile (b) under MQL condition [35].

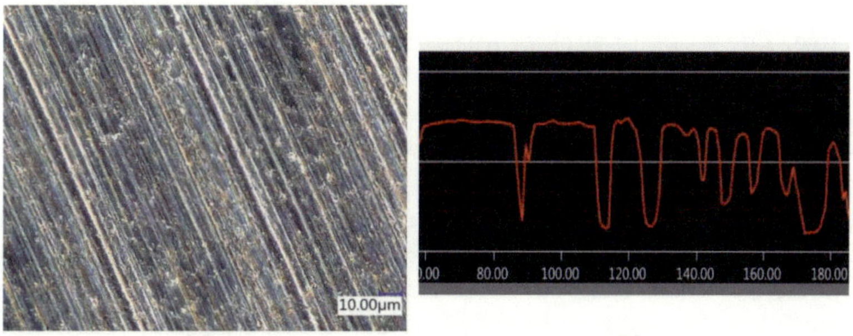

(a) (b)

Figure 20.
Surface microstructure (a) and profile (b) under MQCL condition using pure emulsion-based fluid [35].

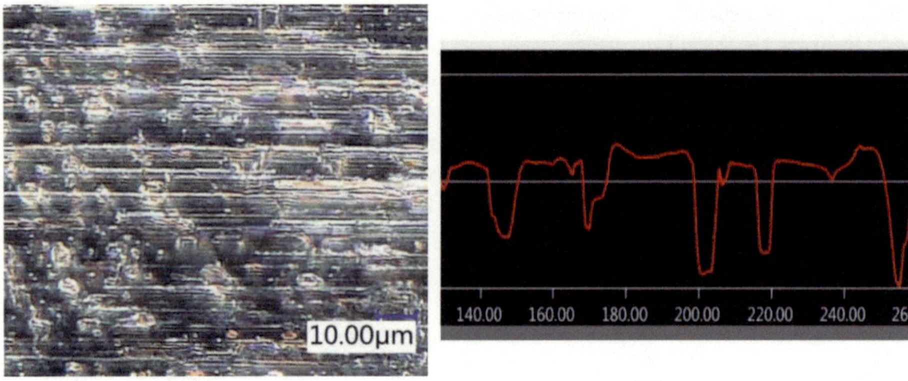

(a) (b)

Figure 21.
Surface microstructure (a) and profile (b) under MQCL condition using emulsion-based nanofluid of MoS_2 0.2 wt% [35].

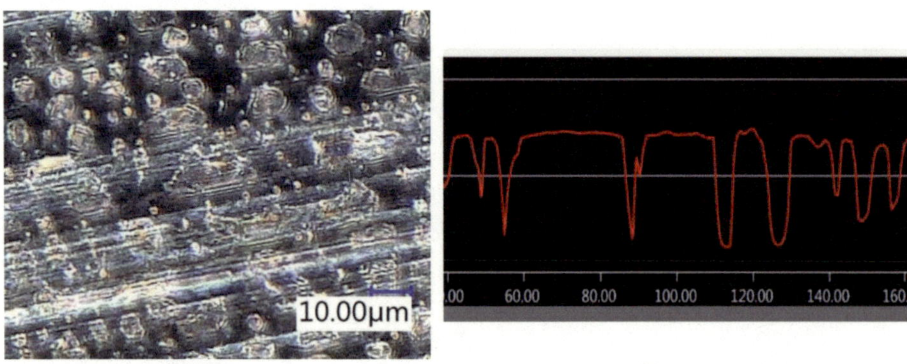

(a) (b)

Figure 22.
Surface microstructure (a) and profile (b) under MQCL condition using emulsion-based nanofluid of MoS_2 0.5 wt% [35].

4. MQCL hard machining using nanofluids

The nanofluids used for MQL are successfully proven to be an alternative solution for difficult-to-cut materials while maintaining its environmental friendly property. Based on this idea, the use of nano additives in MQCL method will bring out the promising solution to increase the hard machining performance. The study of surface quality under MQCL using MoS_2 nanofluid for hard milling is the first attempt to investigate the cooling and lubricating effects [35]. From the obtained results, the values of surface roughness R_a under MQCL using nanofluids are lower than those of dry and MQL conditions. By using MoS_2 nanoparticle concentration of 0.2 and 0.5 wt%, surface roughness is even slightly better than that of MQCL with pure fluid, but the R_a value rapidly increases when increasing the concentration to 0.8 wt%. It can be explained that the morphology of MoS_2 nanoparticles is ellipsoidal with the low friction coefficient up to 0.03–0.05 or even lower [36], by which the better lubricating effect contributes to improve the surface quality. On the other hand, nanoparticles possess the large surface area, which remain on the machined surface to form a thin protective film, which amplifies with the increase of MoS_2 nanoparticle concentration [37]. Furthermore, it also contributes to form MoS_2 tribo-film easily [29], which can be observed from the so-called microbubbles on the machined surface (**Figures 21–22**). The protective film reduces and disappears when increasing the concentration to 0.8%, which causes the negative effect on surface quality [37]. Moreover, the white layer and burn marks are much reduced due to superior cooling and lubricating performance under MQCL condition using nanofluid. From those, the hard machining ability of normal carbide tools improves significantly and is about 157% higher than manufacturer's recommendations [38, 39]. It is the most outstanding finding of this research, and also the proper MoS_2 nanoparticle concentration in emulsion-based fluid was reported about 0.2 and 0.5 wt%, which provides a very important technical guide for further researches and manufacturers. More investigations are necessary to be made for building up technical guidelines and optimizing nanofluid parameters.

5. Conclusion

The application of nanofluids continues to receive growing attention in basic science and machining technology. As shown, nano additives in based fluid of MQL and MQCL methods improve the cooling and lubricating effects as well as tribological property, thus increasing the cutting performance, especially for difficult-to-cut materials. It brings out the alternative solutions for improving productivity and reducing manufacturing cost. From those, the applicability MQL technique having environmental friendly characteristic has been enlarged in hard machining. Furthermore, the use of different types of vegetable oils can fulfill the cooling and lubricating performance by suspending nanoparticles, which is an interesting research topic and exhibits very promising results. On the other hand, MQCL has been considered as another approach for MQL development to overcome the low cooling effect. In this chapter, Ranque-Hilsch vortex tube, a real cooling method, used for creating cooling effect from ordinary compressed air rather than other gas sources to form MQCL method, is the first attempt applied to hard cutting processes. Also, nano additives enriched in MQCL-based fluids are the latest advances in the field of studying MQL and MQCL techniques. The parameters of nanofluid, such as types and size of nanoparticles, concentration, and based fluid, play a key role in successful applications in metal cutting practice, and more studies are needed to make further development and optimize those variables. Those superior cooling and lubricating methods presented in this chapter will contribute to the solutions to reduce/eliminate the cutting fluids and replace dry and wet conditions. It is suitable for protecting our environment and aims for a sustainable production. In the future work, more attention will be paid on other types of nanoparticles, concentration, and parameters of MQL and MQCL methods.

Acknowledgements

The work presented in this chapter is supported by Thai Nguyen University of Technology, Thai Nguyen University, Vietnam.

Conflict of interest

The authors declare no conflict of interest.

Acronyms and abbreviations

d	depth of cut (mm)
V_c	cutting speed (m/min)
F	feed rate (mm/tooth)
F	cutting force (N)
HSS	high speed steel
hBN	hexagonal boron nitride
CNTs	carbon nanotubes
MQL	minimum quantity lubrication
MQCL	minimum quantity cooling lubrication
NFMQL	nanofluid minimum quantity lubrication
ND	nanodiamond
NF	nanofluid
NFs	nanofluids
NP	nanoparticle
NPs	nanoparticles
ANOVA	analysis of variance

Author details

Tran The Long* and Tran Minh Duc
Department of Manufacturing Engineering, Faculty of Mechanical Engineering, Thai Nguyen University of Technology, Vietnam

*Address all correspondence to: tranthelong@tnut.edu.vn; tranthelong90@gmail.com

References

[1] Byrne G, Dornfeld D, Denkena B. Advancing cutting technology. CIRP Annals. 2003;**52**(2):483-507

[2] Zhang K, Deng J, Meng R, Gao P, Yue H. Effect of nano-scale textures on cutting performance of WC/Co-based Ti55Al45N coated tools in dry cutting. International Journal of Refractory Metals and Hard Materials. 2015;**51**:35-49

[3] Paulo DJ. Machining of Hard Materials. London: Springer-Verlag London Limited; 2011

[4] Kumar CS, Patel SK. Effect of WEDM surface texturing on Al2O3/TiCN composite ceramic tools in dry cutting of hardened steel. Ceramics International. 2018;**44**:2510-2523

[5] Su Y, Li Z, Li L, Wang J, Gao H, Wang G. Cutting performance of micro-textured polycrystalline diamond tool in dry cutting. Journal of Manufacturing Processes. 2017;**27**:1-7

[6] S N, G.L S. Drilling performance of micro textured tools under dry, wet and MQL condition. Journal of Manufacturing Processes. 2018;**32**:254-268. DOI: 10.1016/j.jmapro.2018.02.012

[7] Minh DT, The LT. Investigation of MQL-employed hard-milling process of S60C steel using coated-cemented carbide tools. Journal of Mechanics Engineering and Automation. 2016;**6**:128-132

[8] Rahim EA, Sasahara H. A study of the effect of palm oil as MQL lubricant on high speed drilling of titanium alloys. Tribology International. 2011;**44**:309-317

[9] Duc TM, Long TT, Ngoc TB. Performance of Al_2O_3 nanofluids in minimum quantity lubrication in hard milling of $60Si_2Mn$ steel using cemented carbide tools. Advances in Mechanical Engineering. 2017;**9**:1-9

[10] Long TT, Duc TM. Micro/Nanofluids in Sustainable Machining. In: Microfluidics and Nanofluidics. London, UK: IntechOpen; 2018

[11] Lee P-H, Nam JS, Li C, Lee SW. An experimental study on micro-grinding process with nanofluid minimum quantity lubrication (MQL). International Journal of Precision Engineering and Manufacturing. 2012;**13**:331-338

[12] Li B, Li C, Zhang Y, Wang Y, Jia D, Yang M, et al. Heat transfer performance of MQL grinding with different nanofluids for Ni-based alloys using vegetable oil. Journal of Cleaner Production. 2017;**154**:1-11

[13] Ali MKA, Hou X, Mai L, Cai Q, Turkson RF, Chen B. Improving the tribological characteristics of piston ring assembly in automotive engines using Al_2O_3 and TiO_2 nanomaterials as nano-lubricant additives. Tribology International. 2016;**103**:540-554

[14] Pashmforoush F, Bagherinia RD. Influence of water-based copper nanofluid on wheel loading and surface roughness during grinding of Inconel 738 superalloy. Journal of Cleaner Production. 2018;**178**:363-372

[15] Uysal A, Demiren F, Altan E. Applying minimum quantity lubrication (MQL) method on milling of martensitic stainless steel by using nano MoS_2 reinforced vegetable cutting fluid. Procedia - Social and Behavioral Sciences. 2015;**195**:2742-2747

[16] Duc TM, Long TT, Ngoc TB. Effectiveness of alumina nanofluid on slotting end milling performance of SKD 11 tool steel. Journal of Computational and Applied Research in Mechanical Engineering. 2019. DOI: 10.22061/JCARME.2019.4041.1484

[17] ISO 8688-2:1989 (en). Tool life testing in milling — Part 2: End milling; 1989. pp. 9-14

[18] Duc TM, Dong PQ. Influence of coolant fluid and method on tool wear, tool life and the surface roughness in slot machining using end mills. Journal of Science and Technology Technical Universities. 2008;**65**:55-58

[19] Garg A, Sarma S, Panda B, Zhang J, Gao L. Study of effect of nanofluid concentration on response characteristics of machining process for cleaner production. Journal of Cleaner Production. 2016;**135**:476-489

[20] Lee G-J, Park J-J, Lee M-K, Rhee CK. Stable dispersion of nanodiamonds in oil and their tribological properties as lubricant additives. Applied Surface Science. 2017;**415**:24-27

[21] Zhang Y, Li C, Jia D, Li B, Wang Y, Yang M, et al. Experimental study on the effect of nanoparticle concentration on the lubricating property of nanofluids for MQL grinding of Ni-based alloy. Journal of Materials Processing Technology. 2016;**232**:100-115

[22] Luo T, Wei X, Huang X, Huang L, Yang F. Tribological properties of Al_2O_3 nanoparticles as lubricating oil additives. Ceramics International. 2014;**40**:7143-7149

[23] Yıldırım ÇV, Sarıkaya M, Kıvak T, Şirin Ş. The effect of addition of hBN nanoparticles to nanofluid-MQL on tool wear patterns, tool life, roughness and temperature in turning of Ni-based Inconel 625. Tribology International. **2019;*134***:443-456

[24] Duc TM, Long TT, Dong PQ. Effect of the alumina nanofluid concentration on minimum quantity lubrication hard machining for sustainable production. Proceedings of the Institution of Mechanical Engineers, Part C: Journal of Mechanical Engineering Science. 2019;**233**(17):5977-5988. DOI: 10.1177/0954406219861992

[25] Oberg E, Jones F, Horton H, Ryffel H, McCauley C. Machinery's Handbook. 30th Edition, Toolbox ed. United States: Industrial Press; 2016. ISBN: 9780831130916

[26] Duc TM, Long TT, Chien TQ. Performance evaluation of MQL parameters using Al_2O_3 and MoS_2 nanofluids in hard turning 90CrSi steel. Lubricants. 2019;**7**(5):1-17. DOI: 10.3390/lubricants7050040

[27] Maruda RW, Krolczyk GM, Feldshtein E, Nieslony P, Tyliszczak B, Pusavec F. Tool wear characterizations in finish turning of AISI 1045 carbon steel for MQCL conditions. Wear. 2017;**372**:54-67

[28] Maruda RW, Krolczyk GM, Feldshtein E, Pusavec F, Szydlowski M, Legutko S, et al. A study on droplets sizes, their distribution and heat exchange for minimum quantity cooling lubrication (MQCL). International Journal of Machine Tools and Manufacture. 2016;**100**:81-92

[29] Maruda RW, Krolczyk GM, Wojciechowski S, Zak K, Habrat W, Nieslony P. Effects of extreme pressure and anti-wear additives on surface topography and tool wear during MQCL turning of AISI 1045 steel. Journal of Mechanical Science and Technology. 2018;**32**:1585-1591

[30] Maruda RW, Feldshtein E, Legutko S, Krolczyk GM. Research on emulsion mist generation in the conditions of minimum quantity cooling lubrication (MQCL). Tehnicki Vjesnik. 2015;**22**:1213-1218

[31] Maruda R, Krolczyk G, Niesłony P, Krolczyk J, Legutko S, Krolczyk G. Chip

formation zone analysis during the turning of austenitic stainless steel 316L under MQCL cooling condition. Procedia Engineering. 2016;**149**:297-304

[32] Pervaiz S, Deiab I, Rashid A, Nicolescu M. Minimal quantity cooling lubrication in turning of Ti6Al4V: Influence on surface roughness, cutting force and tool wear. Proceedings of the Institution of Mechanical Engineers, Part B: Journal of Engineering Manufacture. 2017;**231**:1542-1558. DOI: 10.1177/0954405415599946

[33] Krolczyk G, Maruda R, Krolczyk J, Nieslony P, Wojciechowski S, Legutko S. Parametric and nonparametric description of the surface topography in the dry and MQCL cutting conditions. Measurement. 2018;**121**:225-239

[34] Hilsch R. The use of the expansion of gases in a centrifugal field as cooling process. The Review of Scientific Instruments. 1947;**18**(2):108-113

[35] Dong PQ, Duc TM, Long TT. Performance evaluation of MQCL hard milling of SKD 11 tool steel using MoS_2 nanofluid. Metals. 2019;**9**:658. DOI: 10. 3390/met9060658

[36] Wang Y, Li C, Zhang Y, Li B, Yang M, Zhang X, et al. Experimental evaluation of the lubrication properties of the wheel/workpiece interface in MQL grinding with different nanofluids. Tribology International. 2016;**99**:198-210

[37] Rahmati B, Sarhan AA, Sayuti M. Morphology of surface generated by end milling AL6061-T6 using molybdenum disulfide (MoS2) nanolubrication in end milling machining. Journal of Cleaner Production. 2014;**66**:685-691

[38] Available from: https://wix.laminatech.ch/img/catalog/1237.pdf

[39] Duc TM, Long TT, Dong PQ, Ngoc TB. Applied research of Nanofluids in MQL to improve hard milling performance of $60Si_2Mn$ steel using carbide tools. American Journal of Mechanical Engineering. 2017;**5**(5):228-233. DOI: 10.12691/ajme-5-5-6